KV-638-350

PAPERS IN EARTH STUDIES

Lovatt Lectures - Worcester

EDITED BY

B.H. ADLAM, C.R. FENN

& L. MORRIS

GEO BOOKS

NORWICH 1982

ISBN 0 86094 117 5

published by Geo Books
Regency House
34 Duke Street
Norwich NR3 3AP
England

Printed by Professional Books Limited

Contents

Editors' Preface

The papers presented in this volume have been drawn from a series of lectures delivered at Worcester during the period 1977 to 1982. The quality of the material presented has always seemed to us to merit the attention of a much wider audience. Accordingly, this collection is published in recognition and appreciation of our 'Lovatt Lecturers'.

No coherent theme is claimed for the volume; rather, the topics embraced reflect the diversity of current research interests in the earth sciences. We are confident that these essays, all of which have been re-modelled for this volume, will provide a useful contribution to the subject areas involved.

B.H. Adlam
C.R. Fenn
L. Morris

Worcester, 1982.

Acknowledgements

The Editors are indebted to the following for permission to reproduce the photographs on the cover:

Worcester floods, December, 1981. by kind permission of the Berrows Organisation Limited, of Hylton Road, Worcester.

Zennor Tors, 1980, by B.H. Adlam.

FOOTNOTE: The Lovatt Lectures were inaugurated in 1977 in recognition of the services of Mr. G.W. Lovatt, former Head of Geography at Worcester and who, as our visitors quickly discover, is very much alive.

1. River power

K.J. Gregory

Fluvial geomorphology has developed rapidly since 1950. This development can be visualised as the cumulative effect of growth in seven areas of activity (Gregory, 1976a), each of which has been subject to internal and external influences. Internal effects have included the paucity of work on fluvial processes by geographers and geomorphologists, so that the growth in fluvial geomorphology occurred in an attempt to redress the balance that had arisen by 1960. External influences have included the way in which progress in hydrology has catalysed development in fluvial geomorphology, particularly through the provision of conceptual ideas, methods and techniques. Although some writers have envisaged a geographical hydrology (Ward, 1978), these hydrological influences were certainly external ones in that their development had been by non-geographers and particularly by engineers. It may be that we have accepted methods and ideas from hydrology without questioning their pertinence to the requirements of fluvial geomorphology. Such uncritical acceptance was essential at first, but now that the research foundation in fluvial geomorphology has been broadened and strengthened, should we be in a position to expect concepts to be modified or to be conceived within fluvial geomorphology itself, as appropriate to the needs of its own research frontier?

In a review of progress made in the study of fluvial processes in Britain, it was argued (Gregory, 1978a) that progress could be resolved into two major components: the first registering the impact of hydrology, and the second providing a prospect for hydrogeomorphology. Hydrogeomorphology may be defined as the study of landforms produced by water, and of the processes responsible, and as such includes both fluvial and coastal domains. In the fluvial domain recent research advances have demonstrated progress in two particular directions. First, there has been a greater emphasis upon temporal change. This is very desirable because it is the only way in which a greater understanding of fluvial processes can be applied to changing environments and to fluvial landforms. This is the underlying theme of S A Schumm (1977) The Fluvial System, and it is a theme echoed in other works on river channel changes (Gregory, 1977a), and on adjustments of the fluvial system (Rhodes and Williams, 1979). A number of research papers published during the last decade have shown how changes of process have featured in specific fluvial systems to induce various types of river metamorphosis. In many cases change has been a response to direct or indirect human influence (eg Gregory, 1979a), but in some cases the anthropogenic effect is inextricably bound up with short term variation in climate. Secondly, there has been a greater awareness of the applied implications of fluvial research. If knowledge of contemporary processes is one ingredient to facilitate the understanding of past temporal change, then why should we not attempt to extrapolate into the future and thereby outline the ways in which fluvial systems may continue to develop? Such extrapolation provides data that can be a useful input to the decision-making process. In considering how applied hydrogeomorphology should become, it was argued (Gregory, 1978b) that there are a range of aspects of the drainage basin and the fluvial system which require investigation in relation to applied problems. The essence of such an applied approach is that it complements work by the hydrologist and by the engineer, and it is achieved by employing spatial and temporal perspectives on meso-scale problems. One must have some awareness of hydraulic engineering, but geomorphic engineering, as envisaged by Coates (1976), is complementary and desirable.

To pursue and to further develop these temporal and applied objectives in research, we may have reached the point at which we need to reconsider our way of looking at fluvial landform. We need a way of developing an index of the channel landform which is effective in that it can be related to processes in the fluvial system. This paper reviews the need for the development of such an index and an approach (Section 1); considers the indices already available (2); proceeds to advocate the use of power and an index expressing it (3); illustrates the index (4); and shows how it may be related to temporal change (5).

1. THE NEED FOR AN INTEGRATED INDEX

Fluvial landforms and the fluvial system have been studied at several levels. Some research has focused on the river channel cross section and upon form and processes within it (in the form of hydraulic geometry, for example); some has concentrated upon the channel pattern and the controls upon several plan form types; and other research has been devoted to drainage networks, either from a topological viewpoint or in terms of the relation between network and climatic input, or between discharge and sediment output. If we add the scale of the bedform in the river channel as studied by the sedimentologist, and at the other extreme, research at the scale of the drainage basin, then we have a hierarchy of five levels at which research investigations have been undertaken: namely, particle, channel cross section, channel planform, drainage network and drainage basin. Although research conducted at any one of the five levels is fully defensible, it is also important to know how any one level relates to the other four. When channel cross section or channel planform is studied, it is notable that there are variations along a reach and through a network. Thus reaches may occur where the relation between channel form and controlling discharges differs from the relationship in reaches immediately upstream or downstream. Perhaps a more pressing need for an integrated appreciation of the relation of the 5 levels arises in the case of the drainage network. Past research has embraced network morphology and topology and also the relations between network character and process. However, many studies have treated the drainage network as a purely spatial pattern and therefore have not allowed for the size of the channels which make up the network - the size which may be visualised as the third dimension of the drainage network. Only by combining the length and size of channels can we really obtain a basis for comparison between areas.

If an integrated index could be developed it could assist further progress in studies of temporal change and in applied aspects of the fluvial system. If change occurs in a fluvial system, then one or more of the several degrees of freedom (Hey, 1978) which the system possesses may adjust. However, only by fully appreciating the inter-relations between the several scales of attention of the fluvial system can we expect to proceed towards a better understanding of which particular degree of freedom will alter in a specific situation. Such an integrated approach may also assist in managing the drainage basin because we now know that the fluvial system should be planned in its entirety and not modified in a piecemeal way irrespective of the consequences elsewhere in the basin. A more specific example of the potential use of an integrated index of the artery of the fluvial system is in its use in flood routing. In models of discharge generation from the drainage basin, the network of channels obviously plays a central role but discharge has often been modelled without reference to channel size. Improved routing models are now beginning to take account of channel dimensions (Gardiner and Gregory, 1982), but an integrated index of the channel network could further assist modelling strategies.

2. INDICES AVAILABLE

Four groups of approaches may be discerned as providing a more integrated appreciation of fluvial landform. <u>Firstly</u>

Figure 1.1. Indices which integrate two or more aspects of drainage basin morphology.

Index	Derivation	Properties integrated	Source
Drainage Factor		Total channel length in relation to basin area; to index "hydrologic shape" of basin	Coulson and Gross (1967) after Sribnyi (1961)
Ruggedness Number	Drainage density (D_d) x Relief (R) $=D_dR$	Length of channel and relief of basin	Strahler (1964)
Stream gradient index	Channel slope at a point and upstream channel length	Slope and channel length. Suggested to reflect stream power or competence	Hack (1973)
Transport efficiency factors	Mean bifurcation ration x total stream length	Channel length and branching of network	Lustig (1965)
	Number of streams of all orders x mean channel slope for basin	Slope and stream order	
Tractive force	τ_0 = Tractive force = γDS where γ = specific weight of fluid D = depth of flow S = channel slope. Depth of flow determined from a modified form of Manning equation.	Discharge/channel size, slope, channel roughness	Graf (1979)

there have been a number of specific studies which are pertinent. In his model of induced erosion and aggradation Strahler (1956) amalgamated consideration of drainage network with channel and flood plain character by showing how erosion upstream was associated with aggradation and flood plain development downstream. Working from the viewpoint of the drainage network, Rzhanitsyn (1963) advocated extending his network analyses to include channel character and hence to relate networks to flow more precisely. In a number of studies, Schumm (eg 1977) has related measures of channel geometry and channel planform and so has integrated these aspects of the system. A further example is provided by Chorley and Kennedy (1971 Fig. 6.26), who portrayed the drainage network as occurring in the range between the two extremes of a basin with one very large channel and, at the other extreme, a basin with a dense network of very small channels. The former provides the most efficient network whereas the latter provides the most probable one. This qualitative approach relates to the notion of minimum energy expenditure referred to below.

A <u>second</u> approach has been accomplished by studies which have utilised integrated indices of drainage basin characteristics. One frequently used approach to the relation between form and process in the fluvial system was to correlate indices of flow or sediment production with a large number of indices of climate and drainage basin characteristics. It was subsequently realised that in view of the correlation between some of the characteristics, and also because of the need to provide a physical explanation of the way in which the characteristics functioned, smaller numbers of more meaningful indices could usefully be employed in a regression model approach. This has sometimes involved the use of indices which compound two or more simpler indices together. Examples of such indices are provided in Figure 1.1. Such compound indices can have disadvantages in that their physical meaning may be difficult to interpret, and specific values of the index can arise in several ways.

These difficulties can be avoided to some extent by a <u>third</u> approach which is based upon the relationship between channel cross section area and total channel length upstream from a specific cross section. This approach has been described in detail (Gregory 1977b). The basis of the technique is illustrated in Figure 1.2 where sections across the river channel may be visualised against the background of a channel network composed of components classified according to frequency of flow. The stages of expansion of the channel network in a particular basin should be related to elements in the compound channel cross section. Although further progress is desirable to establish the precise nature of the relation between different stages of network extent and channel capacity (eg Gregory and Ovenden 1979), the relation between channel capacity to bankfull level and upstream channel length does afford an index which integrates channel cross section and network length and which includes some allowance for planform because of the inclusion of channel length.

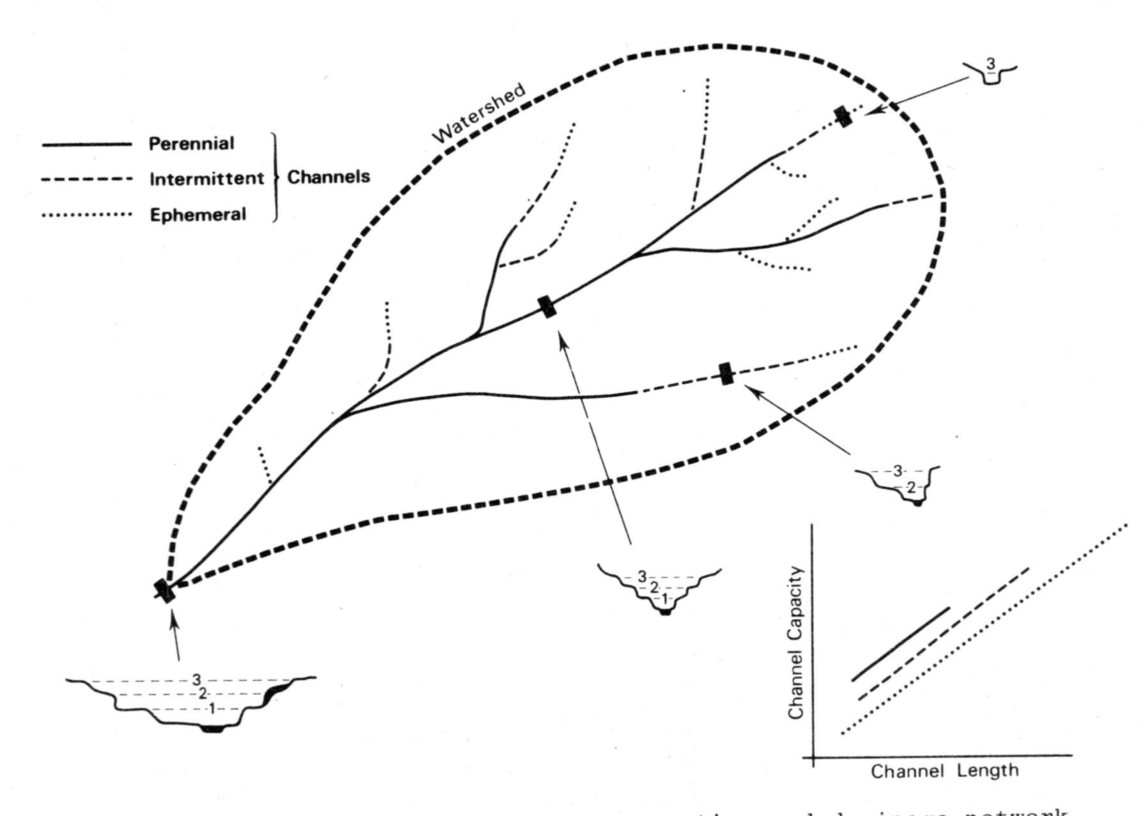

Figure 1.2. Relation between channel cross section and drainage network.

A fourth approach which provides an integrated view of the channel system is based upon minimum variance. Langbein and Leopold (1964) applied least energy expenditure and equal-energy expenditure models to channel geometry and to the river long profile and this approach was followed in the interpretation offered by Chorley and Kennedy (1971). A number of minimization effects have been used in science and in fluvial geomorphology (Williams, 1978) following the approach proposed by Langbein (1964). It has been shown (Yang, Song and Wolden berg, 1981) that a system is in an equilibrium condition when its rate of energy dissipation is at its minimum value. Because the minimum energy expenditure concept applies to several parts of the fluvial system, an integrated approach to the system is implicit.

3. AN INDEX OF POWER

An integrated expression for the morphology of the fluvial system could usefully be based upon power, which is defined simply as the rate of doing work, so that it is equivalent to the amount of work done (W) divided by time interval (t) in the form W/t. If we could characterise fluvial systems according to their potential for producing power then we would have the basis for an extremely useful approach.

Power has been used in other parts of physical geography. Andrews (1972) derived an index of total glacier power for a standard glacier 1 km^2 in size, and this gave a useful means of comparing glaciers. In coastal geomorphology, a classification of coasts can be made according to high, medium and low energy environments. In ecology and biogeography, power has featured significantly through use of comparative estimates of primary productivity and energetics (Odum 1971; Simmons 1978). It has been suggested that power can be used as a concept of wide application in geomorphology (Derbyshire, Gregory and Hails, 1979; Embleton and Thornes, 1979).

To utilise the notion of power in fluvial geomorphology, we require a geomorphological equivalent of the stream power which was advocated by Bagnold (1960) and was developed in relation to sediment transport problems (Bagnold, 1966; Simons, Richardson and Nordin, 1965). Stream power (P) is a function of density of the fluid (d), gravity (g), energy slope (S) and discharge (Q) in the form $P = dgSQ$. In a particular cross section stream power reflects velocity, size of the channel and character of the fluid. A number of studies have recently related minimum stream power to several aspects of fluvial morphology, including the equilibrium geometry of sand bed rivers and river channel patterns (Chang, 1979). A recent proposal (Bull, 1979), has advocated use of a threshold of critical power in streams to discriminate between those situations where lateral cutting occurs and those where alluviation occurs.

This required the specification of critical power as the stream power needed to transport the average sediment load supplied to a stream reach. Ferguson (1981) has drawn attention to the utility of stream power and produced valuable maps of stream power at bankfull discharge for Britsh rivers showing a gradient from high values in the north and west to low ones in the south and east. A further illustration of the potential of power appears from the work of Caine (1976), who estimated the physical work (in joules) represented by different types of sediment movement.

Because the network of stream channels in a basin occupies a central and fundamental role, a meaningful index characterising the essential character of the network could be used to provide an index of potential river power. Water flow in river channels may average approximately 45 cm s^{-1} which is much more rapid than average velocities for other forms of water flow though the drainage basin (saturated throughflow may have a velocity of 20 cm-$hour^{-1}$, pipeflow 10-20 cm s^{-1}, overland flow 25 cm sec^{-1} and source area stormflow 5 cm $hour^{-1}$ (Kirkby, 1978)). Therefore the highest rates of water transport will occur in the stream channels of the drainage network and the longest channels should provide the most rapid water transport, and hence the largest peak discharges. It is not merely the length of channel which is important however because the size of the channels is significant as well. Thus in Figure 1.2 the regression relationship between channel capacity and total channel length could be used to provide an indication of the size of the channel system, because the relation includes both length (ΣL) and size (C) of channel. The regression relation $C = a\Sigma L^b$ therefore tells us something about the character of the network size from the constant (a) which indicates channel size for a unit channel length, and from the exponent (b) which indicates the rate of change of capacity with respect to length. Such regression lines could be used to compare several basins and to indicate potential power of the network, because the larger the network capacity then the greater the velocities and discharges that would be expected. It was suggested that a quantative expression of the size or volume of the drainage network could be obtained from the regression equation by integrating with respect to ΣL between limits of the drainage network (Gregory, 1977b). Therefore for a total channel length of 10 km it would be necessary to integrate the regression equation with respect to $\Sigma L = 0$ to $\Sigma L = 10$ in the form $\int c = a\int_0^{10} \Sigma L^b . d\Sigma L$. This index of volume can provide a quantitative measure which may be compared for several basins. (Ferguson (1979) has questioned the index, but it has been argued (Gregory, 1979c) that his criticisms are not valid). A measure of network volume does not, however, include any measure of relief, and so an index of network power which combines volume of the network (V) and relief (R) was proposed (Gregory, 1979b). This was originally envisaged as either VR or V/R, and although the latter form was advocated (Gregory, 1979b) it may be more effective to use the index in the form VR (Knighton, 1980; Gregory 1980).

An index of network power obtained by combining volume of the network and the relief dimension of the basin can indicate the potential power that the network can induce through the volume of channel available to convey discharge through the network. This approach does have the limitations that it cannot easily apply to very large basins where storage may feature as an important function of the channel system, and the index does not allow for variations in roughness of the channels that could be occasioned by variations in perimeter sediment, vegetation or obstructions. However the index could go part of the way towards providing an index of the physical significance of the drainage network, which would be a useful complement to the indices already available. For example, in a study of channel morphology in southeastern Australia, Nanson and Young (1981) employed a relationship between channel cross-sectional area and drainage area; total channel length could be more valuable as the dependent variable.

4. AN INDEX OF NETWORK POWER AND BASIN PROCESSES

The drainage network is sensitively located between precipitation input to, and discharge output from, the fluvial system. Precipitation acts as one determinant upon the network by influencing the volume of water available. The network also influences the incidence of peak flows in the discharge output from the basin. These two facets of the functional significance of the network can be illustrated by reference to two examples.

In south east Australia there is an increase in average annual precipitation from approximately 850 to 1930 mm over the 200 km from the western part of the New England Tablelands to the coast. As a consequence of this gradient, one would expect a sympathetic response in the drainage network, or in potential river power. To test this possibility, 6 basins were selected (Fig 1.3) extending eastwards over a distance of 130 km and with mean annual precipitation varying from 850 to 1900 mm. Because of the difference in mean annual precipitation, one would expect sympathetic variations in drainage density. However, according to the evidence from 1:25,000 maps supplemented where necessary by information from 1:50,000 planimetric maps and from black and white aerial photographs, there is little variation in drainage density. For the sample areas, the detail shown on the 1:25,000 maps, (all published since 1972), was checked in the field. The principal difficulty was found to be the convention used for the depiction of ephemeral streams. The density of the complete network tends to be lower in the east at 2.85 km km^{-2} than in the west where it is 3.44. The perennial network shows a much greater contrast, because although it is close to 2.85 in the east where there are subtropical conditions with rain throughout the year, it is as little as 0.07 in the west

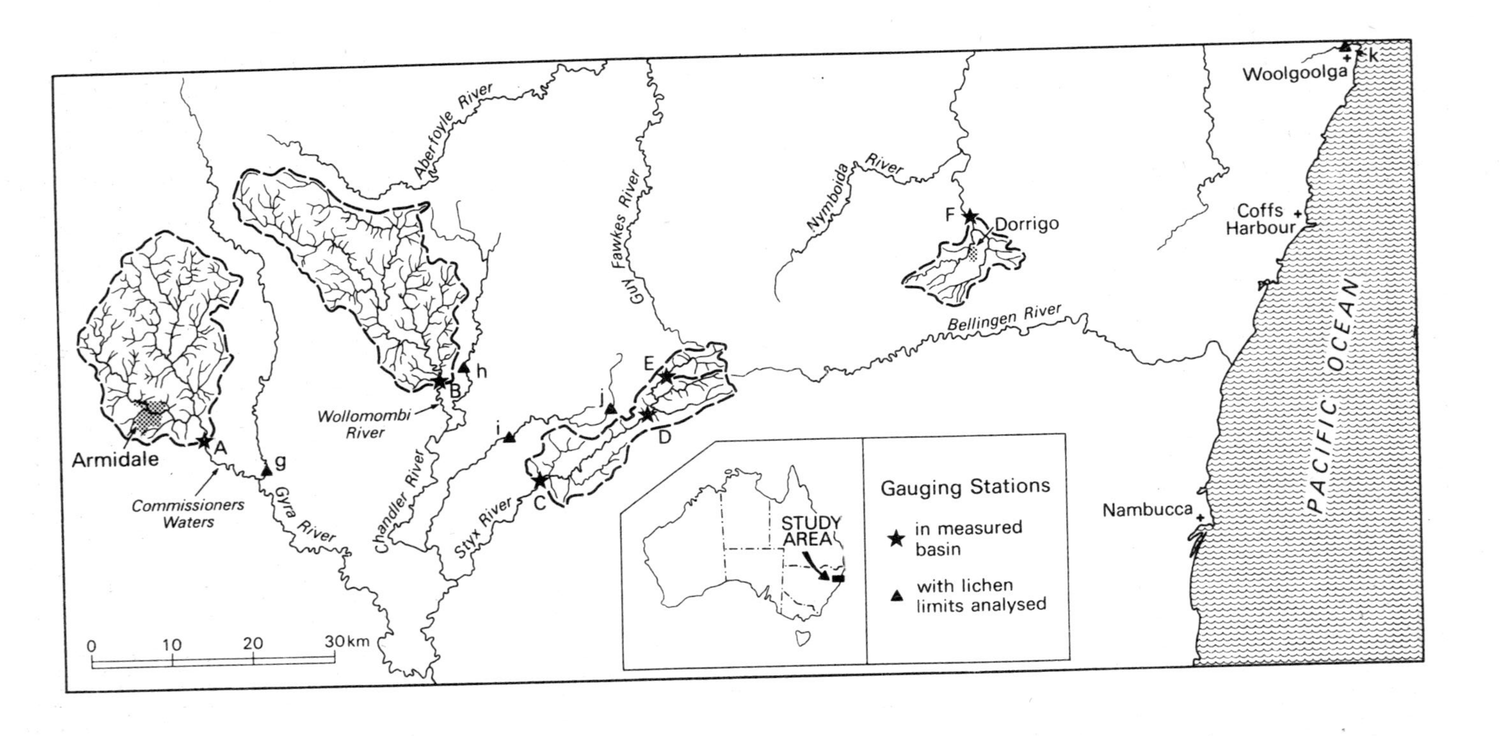

1.3. Drainage basins studied in New South Wales.

where the incidence of storm rainfall is concentrated in brief periods. Peak discharges as well as rainfall amounts are higher in the east and yet densities of the maximum channel network are lower. This anomaly could be explained by variations in channel size, and hence it is necessary to use the regressions between channel capacity and total channel length for the 6 basins, and also to compare the volumes of the channel networks. Channel capacities were surveyed at sites well distributed throughout the six basins (Fig 1.3). A total of 117 sections were surveyed using a quickest level (except in the case of the smallest sections which were surveyed by surveying staff and tape). Channel capacity was determined as exactly possible by reference to sedimentological, morphological and vegetational criteria, and it was found that limits of lichen growth were particularly useful in maintaining consistent definition of channel capacity between sites. All the surveyed sections were plotted, and the measured cross sectional areas to the bankfull level were related to total upstream channel length to provide regression equations.

When using the regression equations to obtain volumes of the channel networks, it is necessary to compare areas of similar size and this was achieved by calculating volumes for basins 100 km^2 in area.

For each basin, the channel length for a drainage area of 100 km^2 was derived from the regression relating total channel length and area. The regression between capacity and total channel length was then integrated between the limits 0 and that channel length appropriate to a drainage area of 100 km^2. The regression relationships demonstrated that channels increased in size relative to channel length towards the east, and the volume estimates indicated that volumes were of the order of two times the size in the east than in the west. If an index of network power is envisaged as the product of the volume and relief of the basin, then because the relief is high in the eastern basins (Fig 1.3) on the escarpment margin of the Tablelands the potential power which can be provided by the networks is greatest in those basins in the eastern area.

This approach therefore affords a means of visualising the potential power of the channel system. Such power may be of fundamental influence upon the generation of peak discharge. It is not simply rainfall amount and rainfall intensity which is responsible for high peak discharges. There must be a channel network which has sufficient potential power to generate the peak flows by giving sufficient length of channel for channel flow at velocities higher than those via other routes through the drainage basin. This Australian example also exposes an outstanding problem of drainage network analysis in relation to process. It has been argued (Gregory, 1976) that the two possible hypotheses (Stoddart, 1969),that drainage density is either directly or inversely related to mean annual precipitation, may be reconciled because the former applies to the perennial network and the latter relates to the complete network of channels including ephemeral streams. However, the way in which the several components of the network relate to the several components in the channel cross section has yet to be fully elucidated and analysed. This is an outstanding problem for future research. The link is required to improve and refine flood routing techniques, and it would be particularly useful in areas with a flashy regime and a network of very variable extent.

A second example is provided by basins in Britain. The Flood Studies Report, published in 1975 (NERC, 1975), utilised existing hydrological data to provide a number of natural and regional relationships which could be the basis for flood estimation at locations anywhere in the United Kingdom. The relationships established necessitated the derivation and use of indices of drainage basin characteristics (Newson, 1978), and a measure of stream frequency was effectively used to characterise the drainage network. Such a measure has limitations, however: it depends upon number of links rather than their length; it was derived from 1:50,000 maps which do not show the complete networks; channel size was not included in the index, or at any other stage of construction of the models. It is therefore desirable to investigate the extent to which channel length and size can be related to discharge for some basins in Britain.

Some constraints had to be imposed upon the selection of basins in Britain. First, basins were selected from areas for which 1:25,000 maps were available, secondly basins not larger than 260 km^2 in drainage area were selected. Thirdly the gauging record from the basins had ideally to have commenced before 1965 to ensure a sufficiently long hydrological record. In addition, it was desirable for the basins to be as homogeneous as possible in their physical character. Twenty-one basins satisfying these criteria were selected to provide as broad a range as possible in rainfall amount and in peak discharge. The majority of basins were selected from western and northern Britain, because the problems of land drainage, channel modification and human influence are so much greater in southern and eastern England. Eastern areas were regarded as appropriate for a subsequent study. In each basin, it was necessary to survey channel cross sections at a range of sites distributed throughout the basin to provide a complete range of channel sizes. The smallest sections could be surveyed with staff and tape whereas the largest were levelled with a quickset level. To minimise variations introduced by the pool and riffle sequence, each profile surveyed was located on the leading edge of a riffle section. For each surveyed cross profile, the likely bankfull level was noted in the field by reference to the distribution of perennial vegetation, and evidence from sediment accumulation and any recent trash lines. Every effort was made to achieve a definition of channel capacity to the bankfull level which was consistent between sites. The surveyed sections were plotted on graph paper and the capacity obtained by measuring the cross sectional area to the bankfull level with a planimeter.

Stream channel length above each site was measured

from available 1:25,000 maps using the second series Regular Edition maps wherever possible. Additional details of the drainage network were obtained from black and white aerial photographs, and from field checking. For each basin, the regression relationship between channel cross sectional area and upstream length of channel was established. The relationship for the Irfon basin is illustrated in Figure 1.4. Comparison of the regression lines for individual basins indicates the nature and extent of the variations present in Britain. For example, the Irfon channels ($C = 0.2364 \Sigma L^{0.9172}$) are consistently larger than those of Halse Water in Somerset ($C = 0.2701 \Sigma L^{0.6663}$). Using the data from some of these basins, it was possible to show (Gregory and Ovenden 1979) how channel network volumes were related to both precipitation amount and to an index of precipitation intensity. Therefore the volume of the networks certainly reflects the amount and rate of input of precipitation. It was also possible to investigate the extent to which the channel networks relate to discharge from the basins. This requires an index of flood flow, and the most appropriate index appeared to be a mean annual flood discharge which was derived from a partial duration rather than from an annual series. The index used was therefore the PT2MAF which is the 2 per year series Mean Annual Flood employed in the Flood Studies Report (NERC, 1975). A discharge of this frequency should be related to flows generated from the complete network, but is not so rare that additional elements would be added to the network. If the volume of channels is related to the index of peak discharge then we have an expression of the relationship. It does not allow for the relief dimension. Thus an index of network power (NP) may be employed as the product of channel volume and relief, which when plotted against the index of peak flow (PT2MAF) gives a relation of the form $PT2MAF = 4.258\ NP^{0.4358}$.

This relationship essentially relates process, as expressed in stream peak discharge, to an index of drainage basin morphology which incorporates size of the stream channels, length of channel and relief of the network. As such the index endeavours to express the volume of the channels comprising the network and the relief through which they fall. Therefore as stream power (W) can be visualised as the product of specific weight of fluid (γ), discharge (Q) and slope (S) in the form $W = \gamma QS$ so we can envisage the channel index obtained as the product of volume and relief as affording an index of network power, because relief will indicate a variable equivalent to slope, and the discharge peak should be controlled by the channel volume (because water velocities will be greater in channels than in other forms of water flow through the basin). Although the index may not be completely satisfactory and other aspects of the network such as roughness should be included (Gregory, 1979b), it is argued that it does help to revise our way of regarding the drainage network (Knighton, 1980; Gregory,1980).

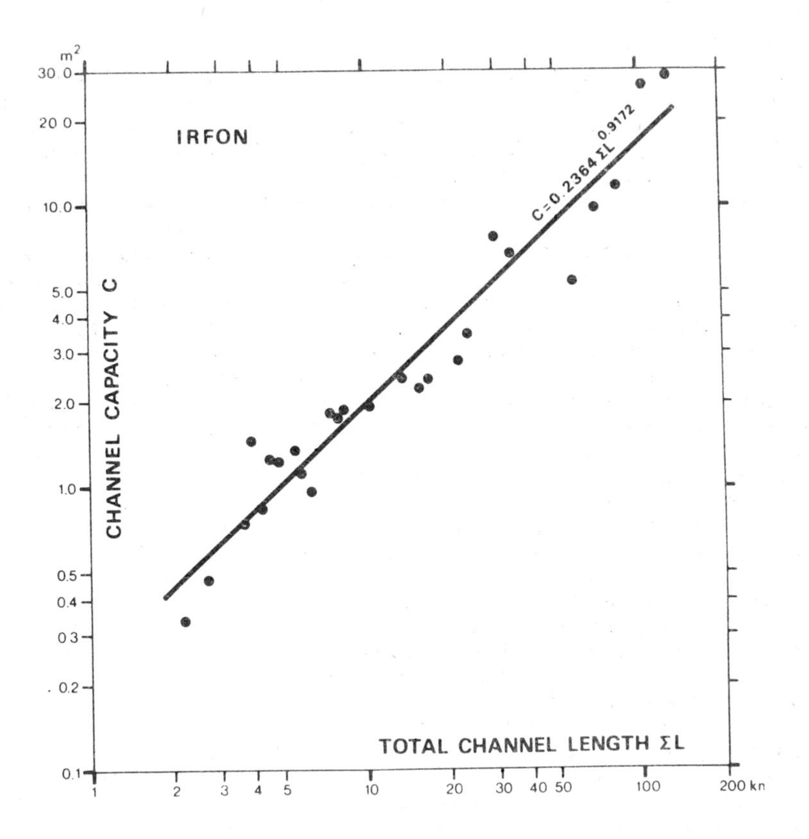

1.4. Example of regression relating channel capacity and channel length.

The Irfon basin has a drainage area of 72.8 km^2 and cross sections were surveyed to the capacity level at 26 sites distributed throughout the basin.

5. CHANGES OF POWER AND TEMPORAL CHANGE

The volume of the channel network is potentially likely to change if there is metamorphosis of the fluvial system. In several studies it has been suggested that the system has a number of degrees of freedom to adjust to a change. These degrees of freedom encompass channel cross-sections, channel planform and slope of the channel, and we could also envisage changes of channel extent or drainage network. These degrees of freedom are therefore contained within the potential power of the drainage network.

River power may be affected by a number of factors. Climatic change may cause variations in the input to the system. Endogenetic movements may significantly effect relief energy changes. Man may interfere with the system in a variety of ways, changing it either directly or indirectly. The major human effects on the fluvial system have been recently reviewed (Gregory, 1979a).

Direct changes include the various techniques of river regulation and channelisation, which may also exert an indirect influence along upstream or downstream sections. Other direct changes may result from deliberate alterations of channel course. The diversion of the Hwang Ho in 1938, when the Kaifeng dyke was deliberately broken in an attempt to use floodwater to stop the advancing Japanese armies, involved major changes of pattern which was not restored until 1947. Interbasin transfers of water may increasingly feature as integral parts of water resource developments (Kellerhalls, Church and Davies, 1979), and these will also have direct and indirect effects.

A variety of less obvious indirect changes in river power have resulted from changes induced by man. Many of these arise as a result of the ways in which man induces changes in discharge, or in water quality, which in turn influence river power and thence fluvial morphology. Examples shown in Figure 1.5 illustrate the influence that reservoir and dam construction, urbanisation, hydroelectric power generation, irrigation and changes of land use throughout the basin have had upon river power and upon the fluvial system. Until the last decade, emphasis in research was placed more upon the direct effects of changes in river power, but now in addition, with a greater awareness of environmental impact, we are able to envisage the extent and nature of indirect, or less obvious effects. An example is provided by dam and reservoir construction. For many years it has been recognised that because of the way in which the reservoir traps sediment there is a likelihood of induced erosion immediately downstream of the dam as the river tries to regain its sediment load. The possibility of such scour has been allowed for during dam design and construction. However a less obvious consequence of dam and reservoir construction has been the way in which the size and frequency of peak discharges and the water quality downstream of the dam have also been modified. Studies have now shown (Petts, 1980) that these changes in water quantity and quality have occasioned changes in the size and character of river channels for considerable distances downstream. When we remember that as much as 15 per cent of flows along European river courses are regulated by dams and reservoirs (Beaumont, 1978) then we begin to see how extensive the consequences of man's effects may be.

6. POWER AND PROSPECTS

Stream power in hydraulics is envisaged as the product of the discharge, the stream slope and the specific gravity of the fluid. If we translate this into more geomorphological terms then we could think of the river power for a basin as the product of discharge, of basin relief and of water quality. This viewpoint can be sustained morphologically by envisaging the potential river power to be reflected in the index obtained as the product of the

Causes	River Power Change	Examples of Morphological result
RIVER DIVERSION	Increase if water diverted into system	Increase in channel capacity, erosion
	Decrease if water diverted out of system	Decrease in channel capacity, deposition
CHANNELIZATION	Increase due to increased water velocity consequent upon reduced channel roughness	Increased channel capacity downstream, erosion downstream
DAM CONSTRUCTION	Power reduced downstream as peak flows are decreased	Decreased channel capacity downstream, aggradation of berms on channel sides
HYDROELECTRIC POWER GENERATION	Power reduced below intake as discharge decreased	Decreased channel capacity downstream. Decrease in size of planform
IRRIGATION	Decrease of power below abstraction points	Decrease in planform and channel capacity
DEFORESTATION	Increase of power arising from increased runoff and higher peak discharges	Increase in channel capacity, change of plan-form, increased network from gullying
URBANISATION	Increased power following urban runoff	Increase in channel capacity, erosion of channels. Aggradation may occur downstream of building activity

Figure 1.5. Changes of river power and possible morphological effects.

network volume and the basin relief. Although detailed variations, such as channel roughness, should also be incorporated, the index may be able to take us some way in the direction of a more integrated understanding of the interrelationship of fluvial forms which is an outstanding requirement for a more hydromorphological approach to the fluvial system. The use of river power as a basic concept in fluvial geomorphology can also be extended to the understanding of changes in the fluvial system. It is through an analysis of such changes in the recent past that we are able to anticipate ways in which fluvial systems could change in the future, and hence to contribute in applications of fluvial geomorphology.

ACKNOWLEDGEMENTS

The author is grateful to NERC for a research grant which supported investigations in British Basins, to the 20th International Geographical Congress fund for a grant towards expenses in Australia, and to Mr W Johns in Australia and to Dr J C Ovenden in Britain who assisted with fieldwork.

REFERENCES

1. Andrews, J.T., 1972, Glacier power, mass balances, velocities and erosion potential, *Zeitschrift fur Geomorphologie Supplement, 13,* 1-17.

2. Bagnold, R.A., 1960, Stream power: A preliminary announcement, *U.S. Geological Survey Circular 421.*

3. Bagnold, R.A., 1966, An approach to the sediment transport problem from general physics, *U.S. Geological Survey Professional Paper 422 I,* 37.

4. Beaumont, P., 1978, Man's impact on river systems: a world wide view, *Area, 10,* 38-41.

5. Bull, W.B., 1979, Threshold of critical power in streams, *Geological Society America Bulletin, 90,* 453-464.

6. Caine, N., 1976, A uniform measure of subaerial erosion, *Geological Society America Bulletin, 87,* 137-140.

7. Chang, H.H., 1979, Minimum stream power and river channel patterns, *Journal of Hydrology, 41,* 303-327.

8. Chorley, R.J., and Kennedy, B.A., 1971, *Physical Geography: A systems approach.* Prentice Hall.

9. Coates, D.R., 1976, (editor), *Geomorphology and Engineering,* State University of New York, Binghampton.

10. Coulson, A., and Gross, P.N., 1967, Measurement of the physical characteristics of drainage basins, *Inland Waters Branch, Department of Energy, Mines and Resources, Technical Bulletin No. 5,* Ottawa.

11. Derbyshire, E., Gregory, K.J., and Hails, J.R., 1979, *Geomorphological Processes,* Butterworths, 310.

12. Embleton, C., and Thornes, J.B., 1979, *Process in Geomorphology*, Arnold, 436.

13. Ferguson, R.I., 1979, Stream network volume: An index of channel morphometry: Discussion and reply, *Geological Society America Bulletin, 90, Part 1*, 606-608.

14. Ferguson, R.I., 1981, Channel form and channel changes, in: *British Rivers*, ed. Lewin, J., 90-125.

15. Graf, W.L., 1979, The development of montane arroyos and gullies, *Earth Surface Processes, 4*, 1-14.

16. Gardiner, V., and Gregory, K.J., 1982, Drainage density in rainfall-runoff modelling, *International Symposium on Rainfall-Runoff Modeling*, ed. Singh, V.P. Mississippi State University.

17. Gregory, K.J.,1976a, Changing drainage basins, *Geographical Journal, 142*, 237-247.

18. Gregory, K.J.,1976b, Drainage networks and climate, in: *Climate and Landforms*, ed. Derbyshire, E, Wiley, 289-315.

19. Gregory, K.J., 1977a, ed., River Channel Changes, *Wiley Interscience*,450.

20. Gregory, K.J., 1977b, Stream network volume: An index of channel morphometry, *Geological Society America Bulletin, 88*, 1075-1080.

21. Gregory, K.J., 1978a, Fluvial processes in British basins, in: *Geomorphology: present problems and future prospects*, eds. Embleton, C., Brunsden, D., and Jones, D.K.C., 40-72.

22. Gregory, K.J., 1978b, Hydrogeomorphology; how applied should we become? *Progress in Physical Geography, 3*, 84-100.

23. Gregory, K.J., 1979a, River Channels in: *Man and Environmental Processes*,123-143.

24. Gregory, K.J., 1979b, Drainage network power, *Water Resources Research, 15*, 775-777.

25. Gregory, K.J., 1979c, Stream network volume: An index of channel morphometry: Discussion and Reply, *Geological Society America Bulletin, 90*, 606-608.

26. Gregory, K.J., 1980, Drainage network power: Discussion and Reply, *Water Resources Research, 16*, 1130.

27. Gregory, K.J., and Ovenden, J.C., 1979, Drainage network volumes and precipitation in Britain, *Transactions Institute British Geographers, NS 4*, 1-11.

28. Hack, J.T., 1973, Stream-profile analysis and stream gradient index, *US Geological Survey Journal of Research, 1*, 421-429.

29. Hey, R.D., 1978, Determinate hydraulic geometry of river channels, *Proceedings American Society Civil Engineers, J.Hyd. Div., 104*, 869-885.

30. Kellerhals, R., Church, M., and Davies, L.B., 1979, Morphological effects of interbasin river diversions, *Canadian Journal Civil Engineering, 6*, 18-31.

31. Kirkby, M.J., 1978, ed., *Hillslope Hydrology*, Wiley.

32. Knighton, A.D., 1980, Drainage network power: Discussion and Reply, *Water Resources Research, 16*, 1129.

33. Langbein, W.B., 1964, Geometry of river channels, *Proceedings American Society Civil Engineers, J. Hyd. Div., 90*, 301-312.

34. Langbein, W.B., and Leopold, L.B., 1964, Quasi-equilibrium states in channel morphology, *American Journal Science, 262*, 782-794.

35. Lustig, L.K., 1965, Sediment yield of the Castaic Watershed, western Los Angeles County, California - a quantitive geomorphic approach, *U.S. Geological Survey Professional Paper 422-F*, F1-F23.

36. Nanson, G.C., and Young, R.W., 1981, Downstream reduction of rural channel size with contrasting urban effects in small coastal streams of southeastern Australia, *Journal of Hydrology, 52*, 239-255.

37. N.E.R.C., 1975, Flood Studies Report 5 Vols.

38. Newson, M.D., 1978, Drainage basin characteristics, their selection, derivation and analysis for a flood study of the British Isles, *Earth Surface Processes, 3*, 277-294.

39. Odum, H.T., 1971, *Environment, Power and Society*, Wiley Interscience, 331.

40. Petts, G.E., 1980, Morphological changes of river channels consequent upon headwater impoundment, *Journal Institution Water Engineers and Scientists, 34*, 374-382.

41. Rhodes, D.D., and Williams, G.P., 1979, *Adjustments of the Fluvial System*, Kendall/Hunt Publishing Co., 372.

42. Rzhanitsyn, N.A., 1963, *Morphological regularities of the structure of the river net.*, Trans. D.B.Kringold, Soil and Water Conservation Research Division, U.S. Dept. Agric. and Water Resources Div. Geol. Surv. U.S. Dept. Interior.

43. Schumm, S.A., 1977, *The Fluvial System*, Wiley, 350.

44. Simmons, I.G., 1978, Physical Geography in Environmental Science, *Geography, 63*, 314-323.

45. Simons, D.B., Richardson, E.V., and Nordin, C.F., 1965, Forms generated by flow in alluvial channels, *Society Economic Palaeontologists and Mineralogists Special Publication 12*, 34-52.

46. Sribnyi, M.F., 1961, Geomorphological characteristics of catchment drainage basins (drainage areas), in: *Problems of river runoff control*, Academy of Sciences USSR. Trans. Israel Program for Scientific Translations.

47. Stoddart, D.R., 1968, Climatic geomorphology: review and assessment, *Progress in Geography, 1*, 160-222.

48. Strahler, A.N., 1956, The nature of induced erosion and aggradation, in: *Man's role in changing the face of the earth*, ed. Thomas, W.L., 621-638.

49. Strahler, A.N., 1964, Quantitive geomorphology of drainage basins and channel networks, in: *Handbook of Applied Hydrology*, ed. Chow, V.T., 4-39 to 4-76.

50. Ward, R.C., 1978, The changing scope of geographical hydrology in Great Britain, *Progress in Physical Geography, 3*, 392-412.

51. Williams, G.P., 1978, Hydraulic geometry of river cross sections - theory of minimum variance, *U.S. Geological Survey Professional Paper 1029.*

52. Yang, C.T., Song, C.C.S., and Woldenberg, M.J., 1981, Hydraulic geometry and minimum rate of energy dissipation, *Water Resources Research, 17*, 1014-1018.

2. British floodplains

J.A. Lewis

Despite their modest relief, floodplains are highly significant landforms. They may provide exceptionally valuable agricultural land and yet be liable to the triple hazards of water-borne pollution, river erosion and flood. Around 8% of the United States is liable to flood, whilst about 40% of Bangladesh is flooded annually following monsoon precipitation. Landforms in such areas generally result from river sedimentation, and such features as levées, point bars, cutoffs and backswamps are described in most geomorphology textbooks. What may be less generally appreciated is that the rates at which floodplain processes operate are highly variable geographically, and that even in Britain the surfaces of some floodplains have been entirely reworked by laterally migrating rivers within a hundred years or so. Other areas, by contrast, appear to have possessed stable rivers with at most only slow overbank sedimentation following floods for a matter of thousands of years. Both such environments have been considerably modified by human activity over the centuries, through canalization or channel stabilization, flood protection, and land drainage works. Such action is almost always expensive, at times ineffective, and certain to produce side-effects in the longer term. Although British floodplains may not be currently liable to changes on the same scale as the arroyo-trenched valleys of the American Southwest, or to such disastrous loss of life as the lower Ganges, it does seem desirable to seek as full an understanding as possible of the processes which operate in this most sensitive of environments.

FLOODPLAIN PROCESSES

Floodplain landforms are generated by fluvial processes in four major ways. The first is through the accretion of channel bed materials, particularly in the form of channel bars. These are of many kinds and have been variously named: N D Smith (1978) listed 32 specific terms preceding the word "bar" and identifying particular bedforms in modern streams! Such bars may be unit forms of simple geometry (e.g. tongue-shaped with a downstream slip-face and an upstream gentle stoss slope) or compounds, both of which may have been modified by erosion before becoming part of the body of floodplain sediments. Bar formation will be important in areas where stream energy upstream of a reach is sufficient to transport bed sediment, where bank erosion and channel movement within a reach leave slack-water voids available for sediment to fill, and where bed sediment is effectively transported into such areas and there deposited. To judge from sediment tracer work, such bed sediment flux in Britain is commonly on a local scale from cut-bank arc to the next available inside bend downstream, on the same side of the stream and opposite a receding cut bank on the other. In some circumstances, bar sedimentation may precede and force cut-bank development as a consequence of channel flow distortion. In others, dispersed bed sediment movement may take place through a series of bends without becoming involved in aggregate depositional forms.

A second process involves sedimentation from suspension either within or beyond river channels. A distinction is commonly made between "lateral" accretion deposits (especially of coarser bed sediment in point bars on the inside of meander bends) and the finer "overbank" sediments deposited by floods. Accumulation of the latter depends on the spatial extent of overbank flows, their frequency, and the sediment concentrations in flooding waters. Actually, finer sediments are also commonly deposited in point bars and in slack-water voids within channels. Flood waters may also occasionally deposit coarse material in overbank locations, so that a sharp distinction between channel and overbank sediments is not nearly so clear as might be anticipated. Overbank sediments should be of most significance where channels are stably located (so that their bed sediment is confined to a narrow ribbon below the channel itself), where there are high suspended sediment concentrations in flooding waters, and where such waters are ponded on flood-plains rather than being channelled or rapidly evacuated from them such that sedimentation is prevented.

A third feature of floodplains consists of the river channels themselves: the main stream, its branches and entrenched tributaries, together with abandoned and cut-off reaches. A great deal of research has been conducted in recent years on active channel forms, not least in Britain. This has been usefully reviewed by Ferguson (1981). As topographic features, or receptacles for overbank sedimentation, dead-channel cutoffs have been somewhat neglected. They may form an important component amongst near-surface floodplain sediments which may be disguised by the thinner blanket of sediment beneath which they are concealed.

Finally, the build-up of sediment on one part of a floodplain (especially as an alluvial ridge along the channel) may lead to the creation of depressions elsewhere. Such sediment occlusion may produce lake, fen or backswamp, whilst depressions may be floored or eventually filled by fine sediment. Chains of lakes or swamps are typical of many floodplain environments, but their existence clearly depends on the propensity for river sedimentation to build up immediately alongside a relatively immobile channel or channel zone, but at a less rapid rate elsewhere.

To these four major groups of fluvial processes must be added colluvial, aeolian, glacial, marine and tectonic processes - because in part many floodplains are features of composite origin. For example 18,000 years ago, during the last (Devensian) glaciation, global sea-level was probably at around -130 m OD.Subsequent sedimentation near present sea level has occurred as a result of an interaction between marine, fluvial and organic sedimentation (see, for example, Greensmith and Tooley, 1982).

FLOODPLAINS AS COMPLEX SEDIMENT STORES

From what has been discussed so far, it may be appreciated that floodplains constitute rather complex sediment stores in which forms develop and sediments accumulate and are preserved for varying lengths of time. What sediments are stacked into storage depend on what particular rivers have available to them, on the ability of rivers to transport such sediments, and on the depositional environments available. The last particularly relate to the nature of lateral channel migration or vertical movement. How long sediments remain in store also depends on continued store-emptying river erosion. The surfaces of some floodplains may be transitory features with a rapid turnover of sediment, whilst elsewhere and in depth floodplains may be underlain by sediments of Pleistocene age. In Britain, such Pleistocene sediments are preserved both in the form of terraces and in alluvial fills extending in places far below the depth of present river activity, so that the broad outline of British floodplains has been inherited from environmental conditions which are known to have differed considerably from those of today. Floodplain sediments indeed are particularly significant in that they preserve a record of past environments and alluvial activity, though one that is far from fully understood. For example, coarse alluvial facies have commonly been interpreted as the deposits of cold-climate braided streams, whereas criteria for braided-stream alluvium are far from straightforward (Bluck, 1979) and the combination of stream process and sediment supply which produces them may not be easily translated into climatic terms. Even for rapidly eroding river banks, the relationship between channel change and the hydraulic and other processes responsible for such changes is far from straightforward (Hooke, 1979; Thorne and Tovey, 1981).

RATES OF FLOODPLAIN REWORKING

The rapidity with which systematic channel changes and floodplain sedimentation can occur may be illustrated by a reach of the River Ystwyth (Fig 2.1). Here the channel was artificially restraightened to run alongside a railway track in 1969, diverging slightly from the line of the track downstream of a footbridge. The photograph shows the channel as it had developed by June 1970. Upstream of the bridge, where the banks are protected and lateral channel movement is prevented, a series of elongated migrating bar forms may be seen, with the low-flow channel overflowing across the bar sides into the deeper scour pools. Downstream of the bridge, a meandering channel has been created. Here no bank protection was provided, bar forms have greater lateral extent and are varied in detail: some have lobate avalanche fronts and some not, there is generally an extensive tail of finer sediment below each bar which was transported around emergent bar crests and deposited at intermediate flows, whilst abandoned slough channels can be seen where bars have become attached to one bank or another. (See Lewin, 1976, for detailed discussion).

On average the rate of lateral channel movement and resedimentation in this example was 18 300 $m^2 km^{-1} yr^{-1}$, or about 50% of channel width in the first year following artificial restraightening. Such change rates are exceptional in Britain and it is helpful to take as long a base as possible by using historical maps and air photographs which record changes over more than 100 years. Using such sources, a random sample of 100 river reaches in Wales and the Borderland showed no discernable shift over many decades on 75% of reaches sampled, with the remaining 25% shifting at between 0.1 and 5.5% of channel width per year (Lewin, Hughes and Blacknell, 1977). More rapid changes have been reported for braided streams which are rare in Britain (Werritty and Ferguson, 1980), with measured mean rates of bank erosion of between 0.08 and 1.18 m yr^{-1} on meandering Devon rivers (Hooke, 1980).

Examination of the geographical distribution of such rates of floodplain reworking shows some discernable pattern. Figure 2.2B shows rates of change at 38 sites in Wales and the Borderland. These are compared with values for gross stream power (Ω) in watts per metre at the same sites (fig 2.2A) calculated as

$$\Omega = \rho g Q S$$

where ρ is 1000 kg m^{-3}, g is 9.8 m s^{-2}, Q is bankfull discharge (m^3 s^{-1}) and S is channel slope (m m^{-1}) (See Ferguson, 1981, for stream power on other Britsh rivers). Generally speaking both streampower and rates of floodplain

Figure 2.1 A reach of the River Ystwyth at Llanilar (SN 628754). Flow is from right to left and the artificially high flow channel upstream of the bridge (about 1/3 of the figure width in from the right) is 36m.

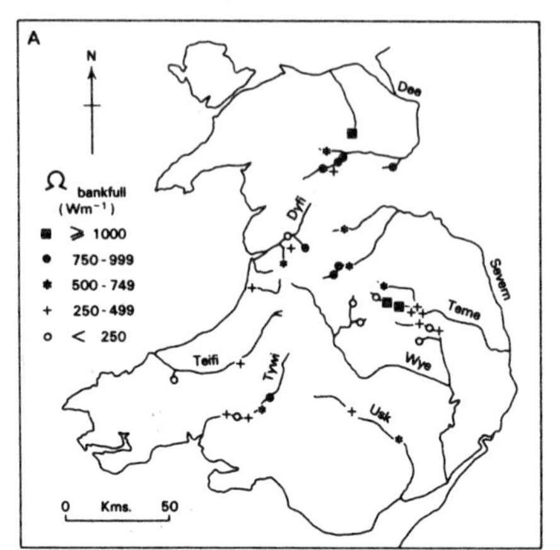

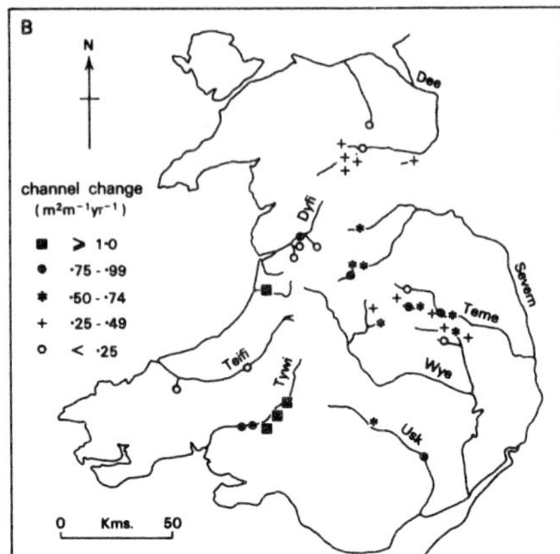

Figure 2.2. Gross bankfull stream power (A) and rates of channel change (B) for 38 actively migrating channel reaches in Wales and the Borderland.

reworking reach a peak in the middle courses of rivers: streampower is less upstream where discharges are very small, and may be less downstream or in lowland rivers where gradients become minimal. However, a plot of gross streampower (or specific power per unit channel width) against change rate reveals a wide scatter. Only a small fraction of the stream power available is actually expended on bank erosion, and it seems likely that this fraction is in fact variable according to channel pattern.

Another way of considering change rates in relation to discharge and slope is shown in figure 2.3. Also plotted is Leopold and Wolman's line, based on extensive data, above which braiding occurs (Leopold and Wolman, 1957). All the streams in this present data set are single thread ones, though some low-sinuosity and rapidly changing ones may be on the braiding threshold. It is also interesting to note that for low slope streams change in dimensionless terms may be greater for small streams than for larger ones, and that there is a general trend in the change-rate data from high rates at steep-gradient small streams (the smallest are omitted) to low rates at low-gradient large ones. This trend is oblique to Leopold and Wolman's

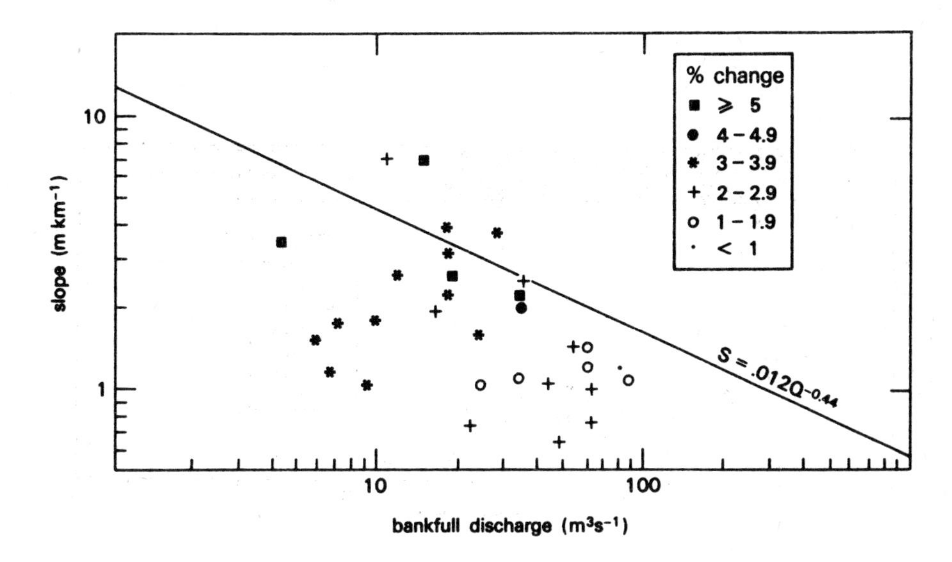

Figure 2.3. Dimensionless channel change rates in relation to channel slope and bankfull discharge for selected river reaches in Wales and the Borderland.

line, and also to lines of equal stream power which are almost parallel.

Figure 2.4 presents change rates in relation to sinuosity and valley slope, and identifies four types of activity pattern, for the same sites that are shown in figure 2.2. It generally shows higher activity rates and greater sinuosities at intermediate slopes, though above slopes of around 4m km^{-1} high rates of change can occur on streams of quite low sinuosity. Though the slope and sinuosity values are different, the pattern is not dissimilar to ones produced experimentally by Schumm and Khan (1972), whilst a downstream sequence from low to high sinuosity streams with accompanying changes in sedimentation pattern has been distinguished by Bluck (1976). These data show that change rates as well as channel patterns are also involved. A four-fold classification of river activity patterns on British rivers, though admittedly somewhat subjective, does help towards an understanding of British floodplain forms.

MOUNTAIN STREAMS

These can be very varied, from low-gradient channels in peat-covered plateaus to steep-gradient boulder-bed channels with waterfalls in bedrock. Upland areas may locally provide large quantities of bed sediment to streams, especially through gullying of Pleistocene sediment, but river channels

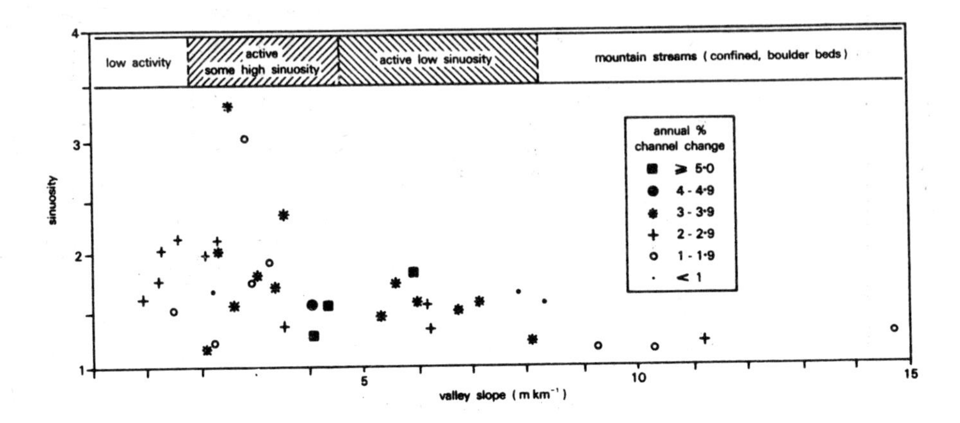

Figure 2.4. Dimensionless change rates in relation to sinuosity and valley slope.

are commonly confined within narrow valleys or between terraces so that floodplain development and channel re-working are both limited (Ferguson, 1981; Newson, 1981). It is interesting that terraces are often underlain by Pleistocene colluvial rather than alluvial sediments. Fluvial facies were probably once present in the valley centre, but these have been removed by subsequent stream action so that the evidence for slope process activity now dominates. Nevertheless, bouldery and peaty floodplains are to be found in steep upland valleys (eg. Lewin and Brindle, 1977; figure 14.4) as well as widespread inherited Pleistocene features.

ACTIVE LOW-SINUOSITY RIVERS

Extensive active braiding in Britain is almost restricted to the Spey river system (Ferguson, 1981), though there are other areas near the margins of upland Britain where low-sinuosity wandering 'gravel' rivers produce complex patterns of floodplain sedimentation. These may be liable to rather rapid channel change and to sediment accumulation in migrating bars that are not stably located on the inside of meander bends. Floodplains in such environments may be complexes of abandoned braid and point bars, together with chutes, sloughs and overbank gravel splays. Figure 2.5 shows a reach of the River Tywi for which on average the channel moves 5.5% of its width per year, and where at any one point, or following an extreme flood, change rates may be much more rapid. Both accretion and sedimentation of coarse and fine sediments occur within and beyond the channel, whilst dead and abandoned channels can also be seen. Such active floodplain environments cause notable management problems, not least because the rate and direction of channel movement at a point are high and unpredictable.

ACTIVE SINUOUS STREAMS

Downstream of the Tywi reach shown in figure 2.5, the channel becomes more regularly sinuous, with bends of repeated geometry and rather less rapid change rate. Patterns of loop evolution involve enlargement and expansion together with cutoff development. Sedimentation is not, however, confined to the inner and downstream margins of meander bends, and complex histories of loops which involve the development of secondary cut-arcs within major loops and sedimentation on the outside of bends are also apparent. The same may be said of other rivers in Wales (Thorne and Lewin, 1979), England (Mosley, 1975; Hooke, 1977) and Scotland (Bluck, 1971), with Bluck's study in particular shedding much light on the way in which floodplain sedimentation occurs.

LOW-ACTIVITY ALLUVIAL STREAMS

As Ferguson (1981) has pointed out, many British lowland rivers are not migrating rapidly at all. This is indicated (and also caused to some extent) by tree-lining of banks, and in lowland England stable underfit channels are very common. Such rivers may transport little coarse material, so that floodplains are built up of silty and clayey sediments, with some organic sedimentation, and coarser sediments only beneath the channels themselves. Figure 2.6 shows cross sections of three lowland valley bottoms in which finer Holocene sediments have accumulated on top of coarser Pleistocene sediments, generally without manifest lateral channel reworking. The Gipping, like other rivers in Eastern England, does not appear to have deposited large volumes of sediments overbank in the Holocene, excepting some recent clays which may relate to a phase of deforestation and ploughing in the catchment. Instead organic sediments have accumulated in this the British

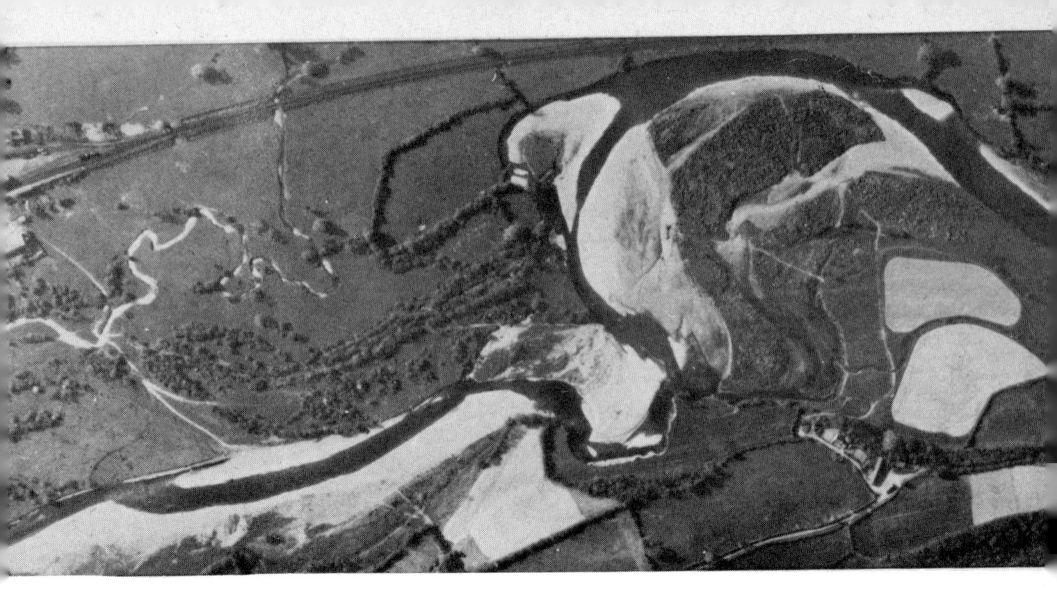

Figure 2.5. A reach of the River Tywi near Llanwrda (SN 720 314). Flow is from right to left.

version of a backswamp environment.

The lower Severn appears to receive a much larger input of fine sediment, and the floodplain soils reflect patterns of sedimentation and surface drainage (figure 2.7). Near the river there are sandy levée sediments which grade to finer materials away from the river, eventually to the gley soils of the Fladbury Series which developed under waterlogged conditions for much of the year. It is interesting to note that all these sediments probably post-date deforestation and accelerated soil erosion elsewhere in the catchment (cf. Shotton, 1978), and that the rather simple levée and floodplain form to the floodplain is indirectly related to human activity.

This brief review of four major categories of river activity in Britain shows that floodplains are expectably of varied geometry and age. Low-sinuosity active streams may rapidly rework their floodplains which have complex relief patterns that are of no great age (several hundred years); inactive lowland streams may have floodplains dominated by overbank sedimentation which has continued for thousands of years, though probably at an accelerated rate since human settlement. Deliberate management and inadvertent side-effects from human activity are widespread, though some of the consequences may be less than generally appreciated.

THE CONSEQUENCES OF HUMAN ACTIVITY

Although increased rates of sedimentation as a result of upstream accelerated erosion seem to have occurred in some areas, river regulation by reservoirs may produce the

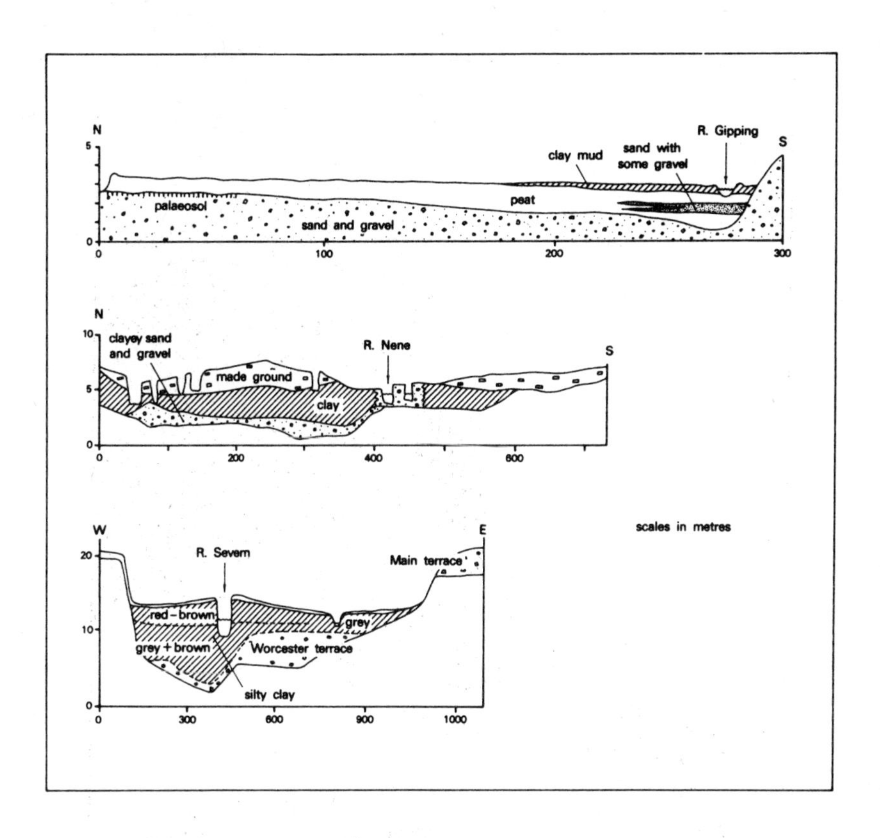

Figure 2.6. Cross sections of three floodplains in lowland England. Top: the Gipping in Suffolk (after Rose, Turner, Coope and Bryan, 1980); Centre: the Nene at Northampton (after Horton, 1970); Bottom: the lower Severn (after Beckinsale and Richardson, 1964).

reverse effect. For example, the regulated river Rheidol has both lower concentrations and yields of suspended sediment, and lower frequency of bed sediment transport, than the adjacent unregulated but otherwise broadly comparable Ystwyth (Grimshaw and Lewin, 1980). In the long run this must affect floodplain sedimentation rates.

Sediment quality has also changed in that many rivers now transport particulate and dissolved pollutants. These may become incorporated into floodplains, though to understand how and where, it is necessary to know how floodplain sedimentation takes place. For example, metal mining wastes became incorporated into floodplains in mid-Wales during the nineteenth century in particular. Metals in

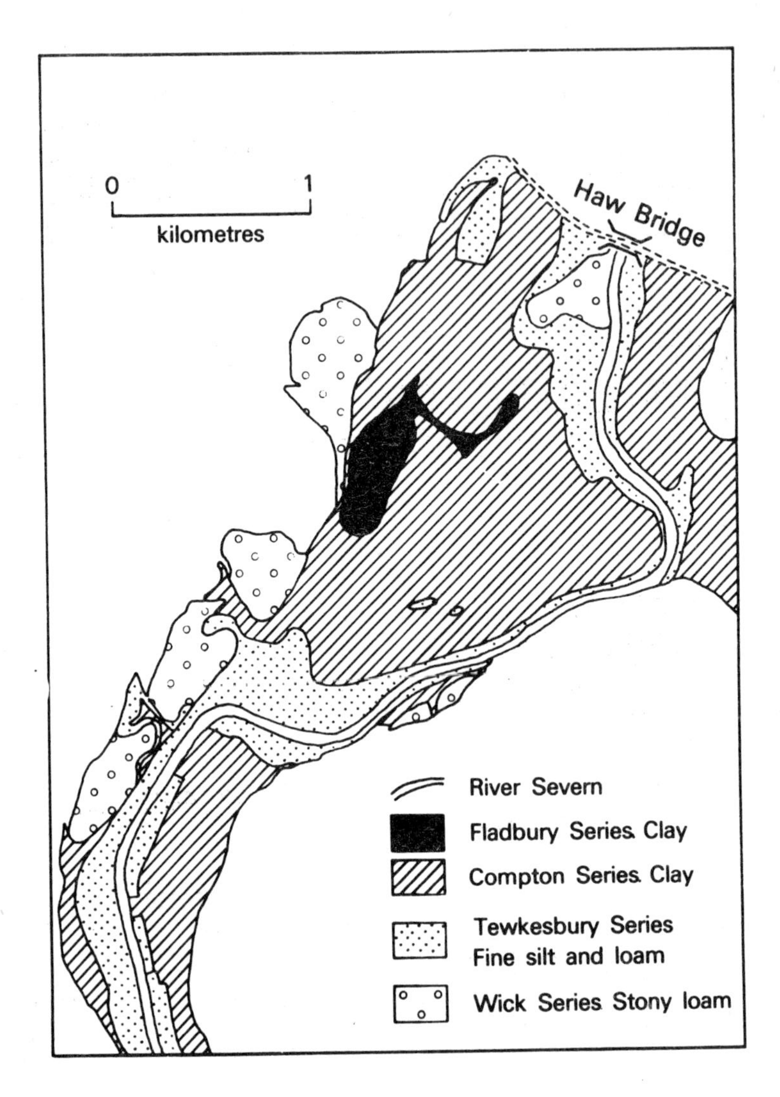

Figure 2.7. Soils of the lower Severn floodplain near Gloucester (SO 8123), simplified from Soil Survey of England and Wales (1973).

sediment form were deposited especially alongside the then channels, particularly in low energy environments where finer materials settled. The result is a spatially variable presence of metals in floodplain soils (Lewin, Davies and Wolfenden, 1977). Work in progress with Dr. B.E. Davies on a length of the floodplain of the River Derwent in Derbyshire (a stable river with a floodplain dominated by overbank sedimentation) suggests that here metals have been added as a more evenly distributed 'top-dressing' across the width of the floodplain.

Both river erosion and flood inundation have been considerably affected by channel modification and river bank protection works (See Gregory, 1979). Channel works have understandably been especially undertaken where natural high rates of channel change would otherwise prevail. This may mean that stream energy is used for other purposes, for example bed scour. Other inadvertent changes may be more subtle. Figure 2.8 shows the floodplain of the lower Severn, including that area mapped in figure 2.6. Since the beginning of the last century, a series of bridges and causeways has been built across the river and valley floor, whilst there is a system of drainage ditches with one-way-flow culverts where they enter the river. The river channel has been enlarged and embanked (with locks).

Even before planned food schemes are put into effect, there is very little likelihood that the passage of flood-water onto and off this floodplain is anything like 'natural', or that the processes of sedimentation and soil development reflected in figure 2.6 are continuing as before. In addition, sediment quality is in any case affected by effluent discharges into the Avon which joins the river at Tewkesbury. Such sediments are known to accumulate on the channel bottom at low flows (Wood, 1981), and may pass onto the floodplain at high ones, so that the benefits of flood nutrient replenishment may be offset. Thus both channel stabilization and flood protection are likely to have consequences in the long term for floodplain development. In fact under natural conditions floodplains form an essential means of extreme flow transmission, perhaps especially so for some streams with cohesive banks and low power (Nanson and Young, 1981) which are adjusted to transmit a larger proportion of their high flows across the floodplain.

CONCLUSIONS

A full understanding of the nature of British floodplains, and the nature of landform changes which occur here, must clearly depend on a full understanding of the whole fluvial system that is involved. This involves processes which operate on different spatial and temporal scales; in one place the rapid and repeated collapse of an eroded channel bank which can be directly observed, and in another the accumulative overbank sedimentation of fine material throughout the last 10,000 years. Human activity may have considerable effects on this system, some unforeseen and some undesirable. Floodplains are extremely sensitive geomorphological environments, and piecemeal management actions and engineering construction are undertaken in a naturally but variably active and hazardous physical environment. A better understanding of such environments is slowly beginning to emerge.

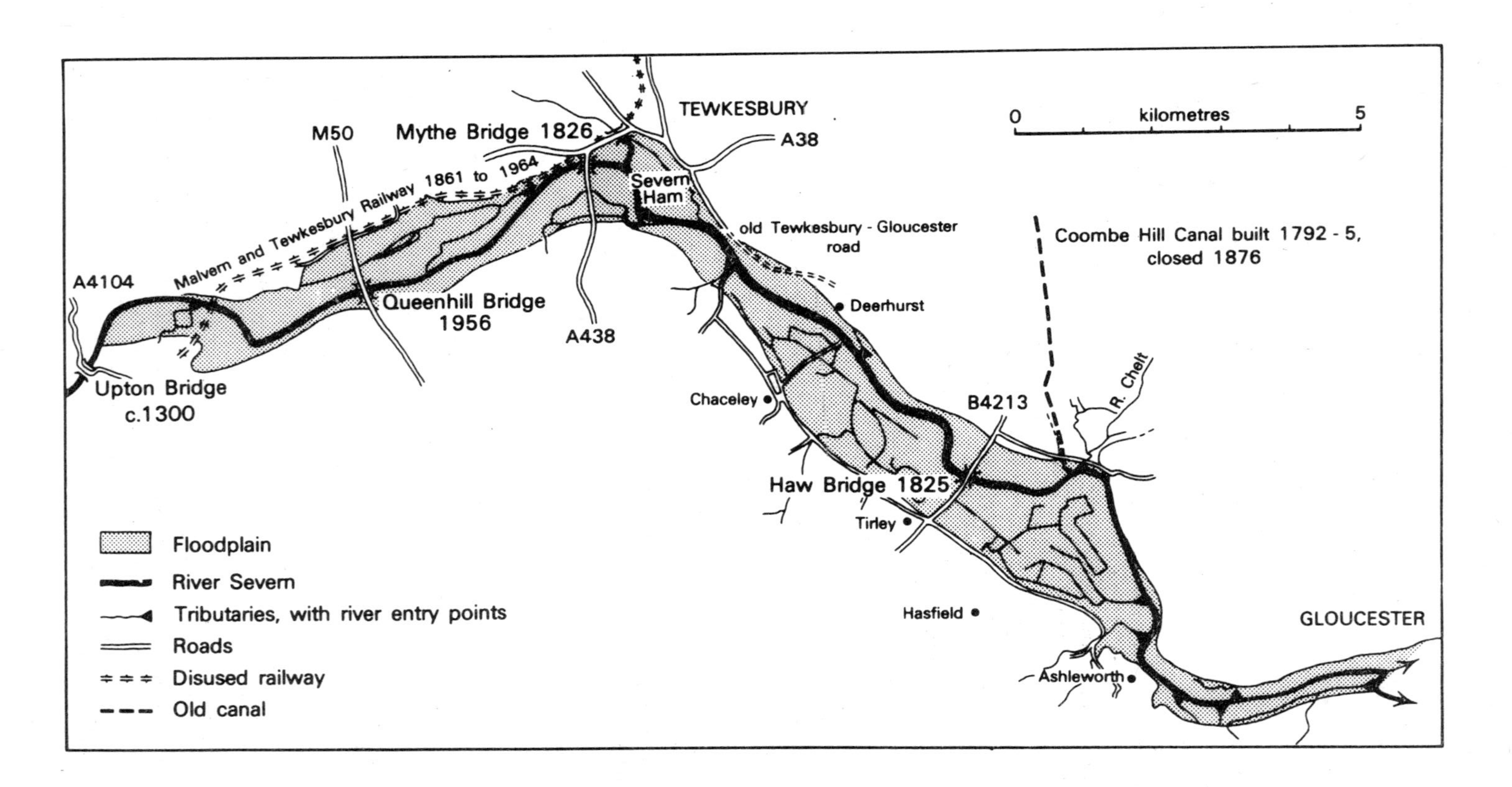

Figure 2.8 The floodplain of the lower Severn between Upton and Gloucester, showing structures and features affecting overbank flows.

ACKNOWLEDGEMENTS

The data in figures 2.2 - 2.5 were obtained through a N.E.R.C. research grant; the author is grateful both to the Council for its support and to D.A. Hughes and C. Blacknell for their considerable assistance.

REFERENCES

1. Beckinsale, R.P., and Richardson, L., 1964, Recent findings on the physical development of the lower Severn valley, *Geographical Journal, 130,* 87-105.

2. Bluck, B.J., 1971, Sedimentation in the meandering river Endrick, *Scottish Journal of Geology, 7,* 93-138.

3. Bluck, B.J., 1976, Sedimentation in some Scottish rivers of low sinuosity, *Transactions of the Royal Society of Edinburgh, 69,* 425-456.

4. Bluck, B.J., 1979, Structure of coarse grained braided stream alluvium, *Transactions of the Royal Society of Edinburgh, 70,* 181-221.

5. Ferguson, R.I., 1981, Channel form and channel changes, in: *British Rivers,* ed. Lewin, J., 90-125, (Allen & Unwin, London).

6. Greensmith, J.T., and Tooley, M.J., eds. 1982, I.G.C.P. Project 61 Sea level movements during the last deglacial hemicycle (about 15 000 years), *Proceedings of the Geologists' Association, 93,* 1-125.

7. Gregory, K.J., 1979, River channels, in: *Man and Environmental Processes,* eds., Gregory, K.J., and Walling, D.E., 123-143, (Dawson, Folkestone).

8. Grimshaw, D.L., and Lewin, J., 1980, Reservoir effects on sediment yield, *Journal of Hydrology, 47,* 163-171.

9. Hooke, J.M., 1977, The distribution and nature of changes in river channel patterns: the example of Devon, in: *River Channel Changes,* ed., Gregory, K.J., 265-280, (Wiley, Chichester).

10. Hooke, J.M., 1979, An analysis of the processes of river bank erosion, *Journal of Hydrology, 42,* 39-62.

11. Hooke, J.M., 1980, Magnitude and distribution of rates of river bank erosion, *Earth Surface Processes, 5,* 143-157.

12. Horton, A., 1970, The drift sequence and subglacial topography in parts of the Ouse and Nene basin, *Report 70/9, Institute of Geological Sciences.*

13. Leopold, L.B., and Wolman, M.G., 1957, River channel patterns - braided, meandering and straight, *United States Geological Survey Professional Paper, 282-B.*

14. Lewin, J., 1976, Initiation of bedforms and meanders in coarse-grained sediment, *Bulletin of the Geological Society of America, 87,* 281-285.

15. Lewin, J., and Brindle, B.J., 1977, Confined meanders in: *River Channel Changes,* ed. Gregory, K.J., 221-223, (Wiley, Chichester).

16. Lewin, J., Davies, B.E., and Wolfenden, P.J., 1977, Interactions between channel change and historic mining sediments, in: *River Channel Changes,* ed. Gregory, K.J., 353-367, (Wiley, Chichester).

17. Lewin, J., Hughes, D., and Blacknell, C., 1977, Incidence of river erosion, *Area 9,* 177-180.

18. Nanson, G.C., and Young, R.W., 1981, Overbank deposition and floodplain formation on small coastal streams of New South Wales, *Zeitschrift für Geomorphologie, NF 25,* 332-347.

19. Newson, M.D., 1981, Mountain streams in: *British Rivers,* ed. Lewin, J., 59-89, (Allen & Unwin, London).

20. Rose, J., Turner, C., Coope, G.R., and Bryan, M.D., 1980, Channel change in a lowland catchment over the past 13 000 years, in: *Timescales in Geomorphology,* eds. Cullingford, R.A., Davidson, D.A., and Lewin, J., 159-175, (Wiley, Chichester).

21. Schumm, S.A., and Khan, H.R., 1972, Experimental study of channel patterns, *Bulletin of the Geological Society of America, 83,* 1755-70.

22. Shotton, F.W., 1978, Archeological inferences from the study of alluvium in the lower Severn-Avon valleys, in: *The Effects of Man on the Landscape: The Lowland Zone,* eds. Limbrey, S., and Evans, I.G., 27-32, (C.B.A. Research Report 21, London).

23. Soil Survey of England and Wales, 1973, *1:25 000 Sheet SO 82,* Norton.

24. Thorne, C.R., and Lewin, J., 1979, Bank processes, bed material movement and planform development in a meandering river, in: *Adjustments of the Fluvial System,* eds. Rhodes, D.D., and Williams, G.P., 117-137, (Dubuque, Iowa).

25. Thorne, C.R., and Tovey, N.K., 1981, Stability of composite river banks, *Earth Surface Processes, 6,* 469-484.

26. Werritty, A., and Ferguson, R.I., 1980, Pattern changes in a Scottish braided river over 1, 30 and 200 years, in: *Timescales in Geomorphology,* eds. Cullingford, R.A., Davidson, D.A., and Lewin, J., 221-233, (Wiley, Chichester).

27. Wood, T.R., 1981, The lower Severn, in: *Field Guides to Modern and Ancient Fluvial Systems in Britain and Spain,* ed. Elliot, T., 6.15-6.20, (International Fluvial Conference, Keele).

3. Karst geomorphology

S.T. Trudgill

The work of a process geomorphologist involves the detailed study of the various components of a system, thus discovering the nature of present processes, and then attempting to relate this to the form and character of existing landforms. The measurement of process is rarely simple, but at least in limestone country it is possible to 'get inside' the system in caves and thus assess patterns of the system's solute load: in limestone regions the water can be examined as it goes into the rock, at various stages within the rock and at the resurgences. Using EDTA titration techniques the solute load can be measured at various points within the system and solute budgets can be derived. Much information of this kind is now available. However, there are two related problems which limit the geomorphological application of such work. Firstly while a key to the understanding of process may be seen as the study of reactions and responses at the micro scale (e.g. Trudgill & Watts, 1979), the results from such work may not scale up to larger scale landforms. Secondly, it is uncertain to what extent the results of the study of present processes may advance our state of knowledge concerning the evolution of landforms which have evolved over a very long historical time-scale.

The problems can be illustrated through a number of examples. For instance the chemistry of the solution process can be studied. It should be emphasised that the process is one of hydrolysis, with the action of dissociated hydrogen ions being the key factor. These may be produced from the solubilisation and dissociation of carbon dioxide in water or the dissociation of organic acids. The measurement of the potential for weathering is thus best assessed by the measurement of free hydrogen ions; measurements of gaseous carbon dioxide in soil are commonly made but may underpredict weathering potential if organic acids are also present.

Significant differences are found between process and erosion rates at the soil-bedrock interface under acid and alkaline conditions. Bedrock solution is less likely to occur beneath alkaline calcareous soils since the water at the bedrock-soil interface has already percolated through carbonates and is liable to be approaching a chemically saturated state. Conversely, under acid soils the per-

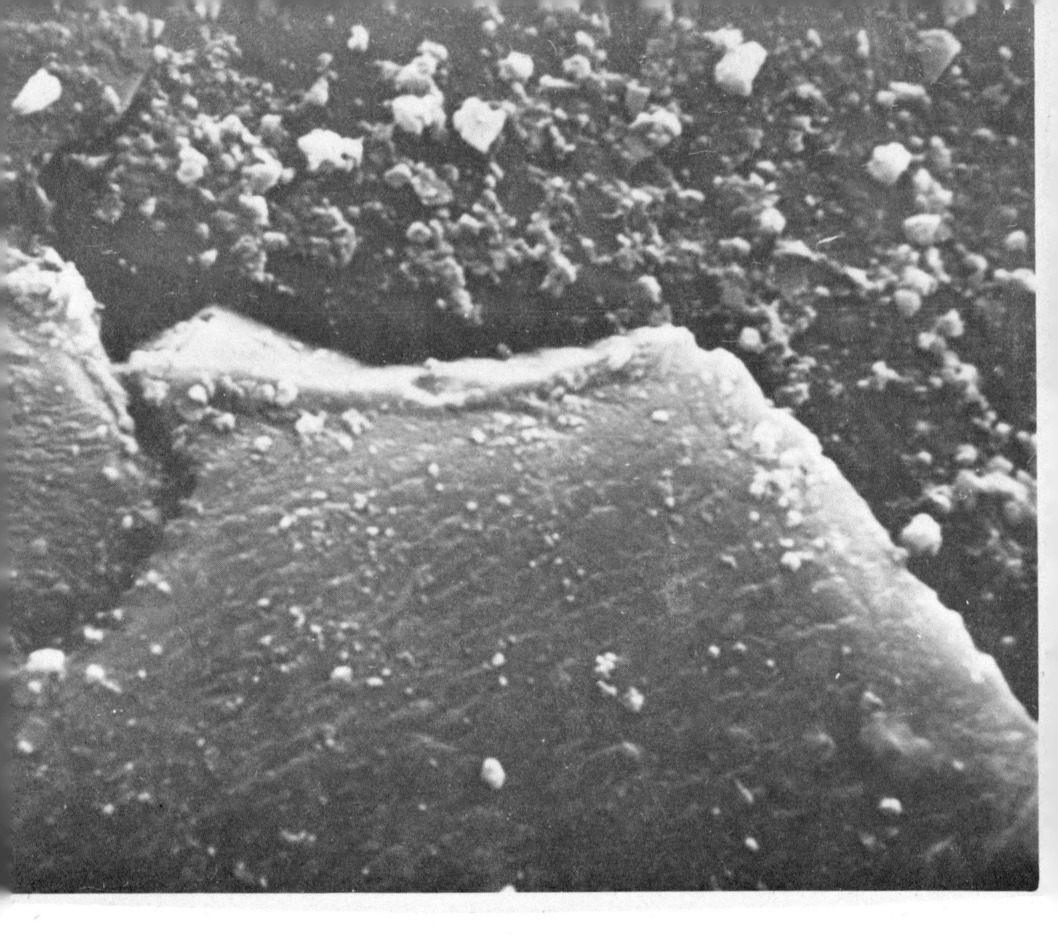

Figure 3.1. S.E.M. of limestone surface from site with acid peat drainage waters. (x 2500).

colating soil waters are under-saturated with respect to carbonates and can attack the rock surface. This effect may be reinforced locally if plants such as *Calluna vulgaris* have become established, as they can acidify the soil by leaf leachates and rhizosphere acidification.

Viewed under an electron microscope, fragments from limestone surfaces in different environments may vary markedly. Surfaces produced under conditions of high acidity (transport limited reactions) tend to be dissolved evenly giving a smooth, polished surface (Figure 3.1). Under conditions of low acidity (rate limited reactions) differential solution can occur (Figure 3.2). The latter tends to take place under soils and the former where organic-rich, acid peat waters drain over limestone. Laboratory control experiments using various acids have suggested that organic acids are not a prerequisite for the production of the smooth surfaces found near peat however. Smooth surfaces are a feature of strong acidity, i.e. a high concentration of hydrogen ions (Trudgill, 1979a).

While the potential for solution processes can be studied using chemical analyses, actual erosion rates can be measured by various techniques. The micro erosion meter is one such technique and is a tripod with micro-meter probe positioned on reference studs set into the rock surface;

Figure 3.2. S.E.M. of limestone surface from an acid (pH 6.5) sub soil site. (x 2500).

(High and Hanna, 1970). The choice of site is often difficult in practice. A site on a smooth rock surface is simple to install but likely to give the lowest results because undissected surfaces may be preserved glacial or other surfaces which are essentially unweathered and unaltered since their inception. Conversely a dissected surface may be the product of more rapid and measurable rates of erosion, but a micro erosion meter site is much more difficult to install on such a surface. Measurement by the meter gives a surface lowering rate for the measurement period. This rate is of great value when comparing one site with another, say with comparisons between rock bands, but valid extrapolations over long periods of time cannot often be made since variations in past conditions are usually unknown.

Weathering rates can also be studied by the use of limestone rock tablets, (Trudgill, 1975, 1977a). Small tablets of limestone can be prepared and placed in a variety of sites. Measurement of weight of tablet at the time of insertion and again when collected produces a weight loss value which may be interpreted as an erosion rate. Again there is merit in comparing one site with another but the extrapolation of rates from present conditions may be invalid.

When discussing erosion rates and process one major question is yet to be resolved; how significant are micro-scale processes in terms of the development of large-scale landforms?

It is certainly apparent that one can relate process to form at the micro-scale but at present it is difficult to extrapolate to large-scale landforms without entering realms of 'geomythology'. At a middle scale, however, the attempt may well be justified.

Limestone pavements are well known features of limestone country and have received a great deal of attention in the literature, (Williams, 1966). It was suggested by Piggott (1962) that a glacial scouring is required to remove the regolith in order to expose the rock in a pavement form. Under a protective cover of calcareous till, glacially scoured surfaces have been preserved intact, weathering being located in the topsoil. Joints are only opened away from the till cover. It can be inferred that in these cases joint opening is thus a post-glacial feature. Observations in recently glaciated alpine areas tend to confirm this. This argument is, however, complicated by the fact that extrapolation from current alpine glaciers to Pleistocene ice sheets may not be valid. On pavements where acid soil covers exist, deep pits are often found under the soils, and indeed, the opening of joints under the soil may result in subsurface soil abstraction - i.e. the soils appear to be 'digging their own graves', (Trudgill 1976a). Soil loss may also be reinforced or entirely caused by the grazing of animals. This reduces the vegetation cover and thus facilitates soil erosion. Where this occurs the pavement which is exhumed shows a dissected surface in marked contrast to the subaerially evolved surface. Under acid till covers dissection and measured erosion rates are both greater than on subaerially exposed pavements.

One way of calculating the age of exposure of limestone pavements is based upon the rate of colonisation of bare limestone by lichens. A lichonometric growth curve may be constructed by the measurement of the diameter of the lichens on gravestones and buildings. The resulting curve may be extrapolated to date the unknown surfaces. Despite all the risks inherent in such an extrapolation, there are some interesting conclusions which suggest that many pavements may have been exhumed from under a soil cover in the last 50-100 years. These are dissected pavements of the type associated with an acid soil cover. Others appear to have been exposed for much longer and are of subdued relief associated with subaerial weathering, (Trudgill, 1979b).

In coastal environments limestone outcrops commonly become increasingly dissected in the middle and lower intertidal zones. This dissection appears to be related to a series of different processes operating at different degrees of coastal exposure and to different levels of tide. The processes are often largely biological. Micro erosion meter measurements have shown that there is a variety of rates of destruction, and the electron microscope has revealed the variety of forms produced by the different organisms. Lichens bore into the rock to produce a honeycomb effect while the effect of blue-green algae (*Cyanophytes*) is to create a lacework form with more holes than rock (Figure 3.3). Winkles (*Littorina* sp.) feeding on the algae leave gouge marks from their rasping radulae while bivalves such as *Hiatella* tunnel into the surface at rates of up to 0.5cm per year. Sea urchins (*Paracentrotus lividus*) occupy hollows which they appear to have excavated, the depths of which seem to increase with the degree of coastal exposure (Figure 3.4). Boring sponges (*Cliona*) are also common on the lower intertidal zone. In the case of the molluscs and sea urchins their shells and tests respectively can be studied to reveal the ages of specimens through their growth rings and approximate rates of boring can be deduced. These may be as high as 1 cm/yr^{-1}. Additionally in a coastal environment the rock will be subjected to the effects of solution by sea water and abrasion by sand and pebbles as well as to the action of storms and salt weathering.

The intertidal zone can be divided into a series of sub-zones each of which has a particular morphology associated with a particular biological process. The greatest dissection occurs where boring algae and echinoderms are prevalent. This is an example of the meso-scale level of process-form study.

Figure 3.3. S.E.M. of an algal-bored intertidal limestone surface, Co. Clare, Ireland. (x 1250).

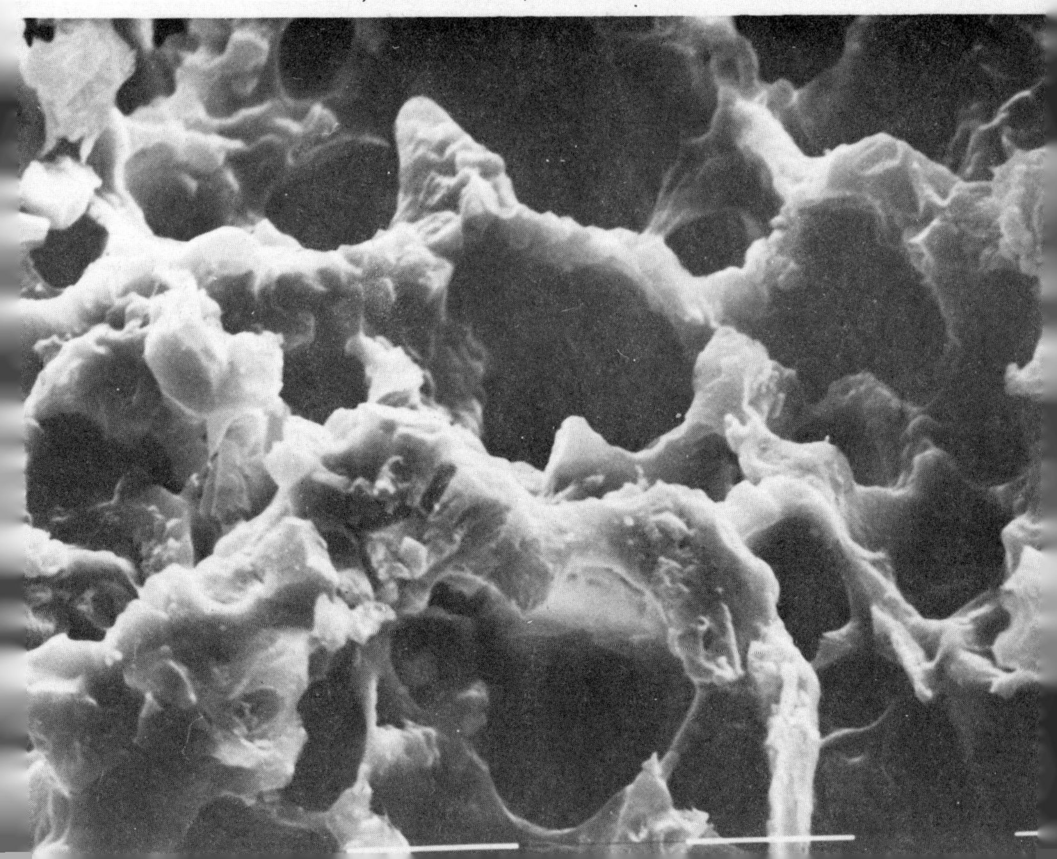

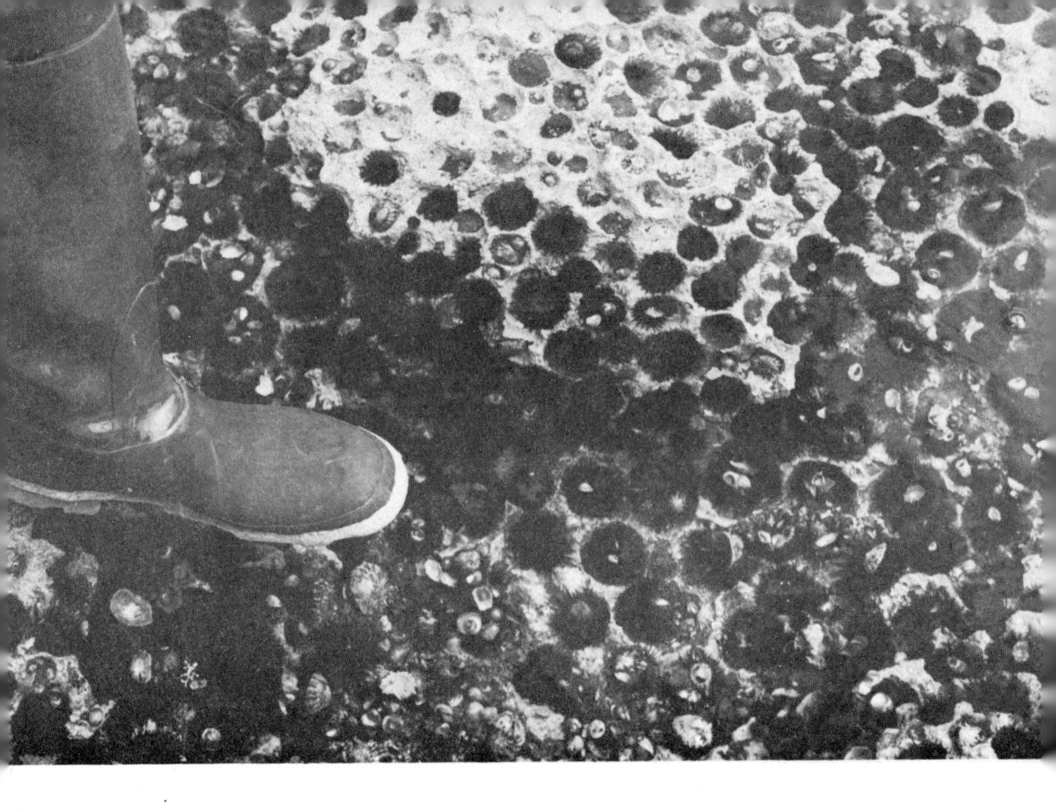

Figure 3.4. Boring Echinoderms, *Paracentrotus lividus*, Co. Clare, Ireland.

The study of process-form relationships on limestones in tropical areas may be seen to be potentially more fruitful in that the extrapolation of present processes and rate measurements may be more valid in regions which have not undergone the geologically recent climatic upheavals of glaciations.

Subaerially exposed surfaces may be deeply dissected and this can be interpreted as being related to the short residence time of water on the rock and rapid rates of reaction. The most rapidly soluble minerals may be preferentially removed. Sub-soil surfaces are fairly smooth, associated with long water residence times and a lack of differential erosion. Measured erosion rates using limestone tablet weight loss, (Trudgill, 1977b) do not correlate strongly with measured soil carbon dioxide levels, as high CO_2 levels are associated with the less permeable soils. Erosion rates correlate well with soil pH however, reinforcing the conclusions about the limitations of carbon dioxide measurements and the usefulness of acidity measurements outlined earlier.

In summary, five main points emerge:

1. It is possible to measure erosion rates for limestone outcrops under different conditions, especially at the micro-scale.
2. It is possible to determine the distribution of erosion rates in different parts of the karst system.
3. It is apparent that the concept of 'climatic geomor-

phology' should be strongly tempered with the study of lithological variations and stability of given conditions (Trudgill, 1976b).

4. Despite and relative ease of measurement of current rates of process, and their distribution within the system, the validity of long-term extrapolation of the data derived is often very limited.
5. Despite the recent advances in process geomorphology, the micro-scale of operation and short-term measurement times make their application to the understanding of large scale landforms difficult.

The relationships between present measurable process studies and the long term evolution of landforms still remains a challenging field of research.

REFERENCES

1. High, C.J., and Hanna, K.K., 1979, A method for the direct measurement of erosion on rock surfaces, *British Geomorphological Research Group, Technical Bulletin, 5*.

2. Piggott, J.D., 1962, Soil formation and development on the Carboniferous Limestone of Derbyshire. Part 1 - Parent Materials, *Journal of Ecology, 50,* 145-156.

3. Trudgill, S.T., 1975, Measurement of erosional weight-loss of rock tablets, *British Geomorphological Research Group, Technical Bulletin, 17,* 13-19.

4. Trudgill, S.T., 1976a, The erosion of limestones under soil, and the long-term stability of soil-vegetation systems on limestone, *Earth Surface Processes, 1,* 31-41.

5. Trudgill, S.T., 1976b, Rock weathering and Climate: Quantitative and Experimental Aspects, in *Geomorphology and Climate*. ed Derbyshire, E. (Wiley, London)

6. Trudgill, S.T., 1977a, Problems in the estimation of short term variations in limestone erosion processes, *Earth Surface Processes, 2,* 251-256.

7. Trudgill, S.T., 1977b, The role of soil cover in limestone weathering, Cockpot Country, Jamaica, *Proceedings of the Seventh International Speleological Congress, Sheffield, British Cave Research Association,* 401-404.

8. Trudgill, S.T., 1979a, Chemical polish of limestone and interactions between calcium and organic matter in peat drainage waters, *Transactions of the British Cave Research Association, 6,* 30-35.

9. Trudgill, S.T., 1979b, The age of exposure of limestone pavements - a pilot lichonometric study in Co. Clare,

Eire, *Transactions of the British Cave Research Association, 6,* 10-14.

10. Trudgill, S.T. and Watts, S.T., 1979, An investigation into the relationship between solvent motion and the solutional erosion of an inclined limestone surface, *Transactions of the British Cave Research Association, 6.*

11. Williams, P.W., 1966, Limestone Pavements, with special reference to W. Ireland, *Transactions of the Institute of British Geographers, 40,* 155-172.

4. Agriculture and the hydrological regime

G.E. Hollis

PREAMBLE

The Resources and Environment Programme of the International Institute for Applied Systems Analysis encompasses a study of management of agricultural watersheds and the impact of agricultural practice on water resources. The Conference held in 1979 and organised by the I.I.A.S.A. and the Czechoslovak Academy of Sciences at Smolenice brought together specialists from twenty nations, and focussed on hydrological and engineering aspects of the subject. This is an unrevised version of a paper presented at that Conference.

INTRODUCTION

A desire for a greater degree of national self-sufficiency in food and timber has lead the U.K. Government to propose the expansion of upland forests and increase in food production through a further intensification of agriculture. The severe drought of 1975-76 focussed the attention of scientists and the public at large onto the problems of water supply and water quality. The reorganisation of water management, to provide large multi-functional water authorities in England and Wales and strengthen River Purification Boards in Scotland, has facilitated an integrated overall view to be taken of the hydrological cycle, its use and management. Against this background, agricultural practises have been the subject of considerable research because of their hypothesized hydrological effects. This paper reviews recent U.K. work on hydrological changes consequent upon upland afforestation, on flood runoff and ecological modifications after improvements in land drainage, and on possible mechanisms to explain a disturbing rise in the concentration of nitrate-nitrogen in surface waters.

UPLAND AFFORESTATION

In 1924 there were 573,000ha of high forest in Great Britain comprising 2.5% of the total land area. By 1976, managed forest land had risen to 1.69 mill.ha which when added to the 300,000ha of unproductive woodland comprised 8.5% of the land area. Most of the new planting has involved coniferous plantations in the uplands, lowland forest being restricted to the old Royal forests or areas of poor sandy or chalky soil afforested in the '20s and '30s (Binns, 1979). This extensive upland afforestation, coupled with Law's (1956) finding that the annual water loss from a Sitka Spruce plantation was 290mm greater than from grassland, has lead to the Institute of Hydrology's Plynlimon experiment in Central Wales reported by Clarke and McCullock (1979).

Two small basins, the 1055ha grass covered Wye catchment and the 870ha Severn catchment with a 66% cover of conifers, are being used for a classic paired catchment study and as a base for process investigations. Thirty eight ground level or canopy level monthly raingauges, six autographic raingauges and eight flow gauging stations, including six specially designed steep stream structures, form the basic instrumentation. In addition automatic weather stations in each major domain record net radiation, total solar radiation, wet bulb depression, wind run and wind direction on magnetic tape every five minutes. Extensive networks of soil moisture access tubes facilitate monthly recording of soil moisture by neutron probe. Snowfall is recorded by terrestrial photogrammetry and snow courses.

Figure 4.1 gives annual totals for rainfall and runoff for each catchment and the difference between water losses from the wooded Severn and grassland Wye. During the years 1970-1 there were problems with sediment accumulation in the Severn flume. The construction of an upstream sediment trap solved the problem and data for 1972-4 "were used to estimate stream flow for the ..... period 1970-1" (Clarke and McCulloch, 1979). The mean value for the excess of losses in the forested Severn compared to the grassland Wye for the last four years, free of the complication of the earlier period, is 281 ± 20mm. This may be an underestimate of the extra losses resultant on having forests. Eighty-nine percent of the annual net radiation is used for evaporation in the Severn basin but only 65% in the Wye which suggests that the extra losses would have been greater if trees were growing on the upper 34% of the Severn catchment. It is hypothesised "that the additional water loss from the forest is the result of evaporation of raindrops intercepted by the tree canopies" (Clarke and McCulloch, 1979) and consequently it is argued that losses

Figure 4.1. Annual Values of Precipitation (P), Streamflow (Q) and P-Q for the Grassland Wye Catchment (W) and Wooded Severn (S) Catchment, 1970-75 Units: mm.

Year	P:		Q:		P-Q:		$P_S-Q_S-P_W+Q_W$
	W:	S:	W:	S:	W:	S:	
1970	2869	2690	2415	(1991)	454	(699)	(245)
1971	1993	1948	1562	(1328)	431	(620)	(189)
1972	2131	2221	1804	1567	327	654	327
1973	2605	2504	2164	1823	442	681	239
1974	2794	2848	2320	2074	474	774	300
1975	2088	2121	1643	1406	456	715	258
MEAN	2415	2388	1985	(1698)	431	(690)	(260)

of a similar magnitude could occur for forests growing in regions of quite different precipitation. These findings have been confirmed by studies of an 84 m^2 lysimeter plot containing 26 Norway Spruce trees, 10m in height and located in the Severn catchment. Calder (1976) found that "the annual loss recorded at the lysimeter, amounting to twice the Penman (open water evaporation rate),demonstrates the magnitude of the evaporation losses that can arise from forests in a high rainfall area". Clarke and Newson (1978) showed that "in the drought years of 1975 and 1976 losses from the Wye catchment under hill pasture were about 20% of annual precipitation, while those for the Severn were rather more than 30% of annual precipitation". They employed a simple simulation of a reservoir to show that failure to allow for a land use change (from grass to forest) could nullify a water management procedure designed to maintain supply even during a drought year.

Detailed studies of the interception process have generally supported these catchment scale findings. Rutter (1963) found interception loss in a Scots Pine plantation in Berkshire to be much higher than Penman estimates of open water evaporation and therefore higher than that to be expected from grass. Subsequently, Rutter et al. (1971, 1975, 1977) developed and validated a mathematical model for interception in a variety of forest types. The model operated on an hourly basis and computed a running water balance for the canopy. The duration of rainfall, as a measure of the time when the canopy was wet, was shown to explain much of the monthly variation in interception loss and constituted an important element in the model. The model has been extended by Calder (1977) to include both transpiration and interception. He was able to test the model over an exceptional range of conditions during 1976. Calder (1979) demonstrated the close correlation between predicted and observed losses for the Plynlimon lysimeter. He showed that, in 32 months spanning 1974-76, transpiration losses from the lysimeter totalled 900mm and interception losses reached 1600mm with a gross precipitation of 5464mm. Work with micrometeorological sensors above Thetford forest in East Anglia (eg. Gash and Stewart, 1977) has been reviewed by Calder (1979). For 1975, rainfall was 595mm; interception loss 213mm; and transpiration 353mm. This suggests a total loss less than would have occurred from grassland inspite of the high rates of evaporation observed when the canopy was wet. Calder (1979) explains this apparently anomalous result by reference to the infrequency of rain; only 350 hours in 1975. This means that the interception process operates for insufficient time to make up the difference between the low transpiration rate from the trees during dry weather and the relatively higher one from grass. Gash (1979) has used the Thetford data "to create a model of interception loss which is simpler and easier to apply than the Rutter model, but ... contains a great deal of the objectivity and physical reasoning behind the model".

The high sediment yield of forested basins has been quantified by Painter et al. (1974). They found that the forested Tanllwyth sub-catchment at Plynlimon had four times the sediment load of the pasture land Cyff basin. Erosion of the drainage ditches, which are cut to assist tree establishment and growth, was largely responsible. Their work on the Coalburn in Northumberland revealed an increase of two orders of magnitude in sediment transport after ditching had taken place. The reduced water yield from afforested land, the high rates of erosion from forest drains and possible water quality problems from fertilizers and herbicides have been recognised by Binns (1979), a Forestry Commission scientist. His review of forestry practises shows that, whilst water quality effects can be rendered negligible and erosion limited, "the effect on water yield over the country as a whole will be determined more by policy on land acquisition and use than by the way the actual operations are done".

LAND DRAINAGE

Land drainage encompasses arterial drainage, the straightening, deepening and embanking of water courses and rivers, and underdrainage, the draining of fields by a combination of some or all of tile or plastic pipes, permeable backfill mole drains or subsoilings (Cole, 1976; Meirs, 1974; Green, 1979). Records of arterial drainage are not centralised however the Water Data Unit (1975 and 1977) have shown that in England and Wales the capital expenditure on land drainage and flood protection (which very largely excludes underdrainage) was £17.04 mill in 1974 and £27.49 mill in 1975 with £23.47 and £30.23 being spent in the respective years on the recurrent account. Figures 4.2 and 4.3 show the extent of underdrainage in the United Kingdom in the last 30 years. Green (1979) has shown that more than 2% of some areas of Eastern England are receiving underdrainage each year in the 1970s. Essex has 15% of its whole area, but 30% of its arable area, underdrained. The prevalence of underdrainage suggests that substantial agricultural benefits accrue as a result. The Field Drainage Experimental Unit has used experimental sites to quantify the benefits of land drainage for watertable control (eg. May and Trafford, 1977), crop yields (eg. Armstrong, 1977) and soil temperature (eg. Waters, 1977). However five years' work at the FDEU site at Abbots Ripton showed "no differences in watertable control or crop yield between the underdrained control and drains at 40 metres with moling (Kellet and Davies, 1977). This section examines the view that land drainage exacerbates flood problems and reports on the lively national debate on the relationship between land drainage and nature conservation.

Hill (1976) has summarised the effects of land drainage but he pointed out the lack of detailed evidence on specific changes. Rycroft and Massey (1975) said that "published evidence that drainage increases flooding is almost non-existent". Howe et al. (1966) demonstrated a significant increase in flooding in the Severn and Wye catchments (their studies used records from the lower reaches, not the Plynlimon headwaters) from 1911-1964. However they could not demonstrate whether this effect was the result of either a demonstrable increase in heavy rainstorms or land management by way of afforestation and land drainage. Green (1973) worked on the Willow Brook near Peterborough, where there is extensive underdrainage. He showed that the number of days that flow exceeded 2.12 cumecs had increased from an average of 4 days per annum for 1945-60 to 15 days per annum for 1966-71. He has also asserted (Green, 1979) that the underdrainage of 30% on the Bury Brook catchment in Cambridgeshire has increased the number of flood peaks compared to the neighbouring Harper's Brook which is only 7% underdrained. These analyses, however, say little about the effect of land drainage on the form or magnitude or significant flood hydrographs. The drainage of the 15km^2 Glenullin catchment has received detailed attention from Wilcock (1979). This boulder clay basin, with extensive raised peat bogs, has had nearly 2kms of channel cleared and scoured alongside one bog and extensive ditches and underdrains installed in a second. Wilcock found that flood flows were reduced in magnitude and frequency during the post-drainage period. Total annual water yields and low to middle range flows were

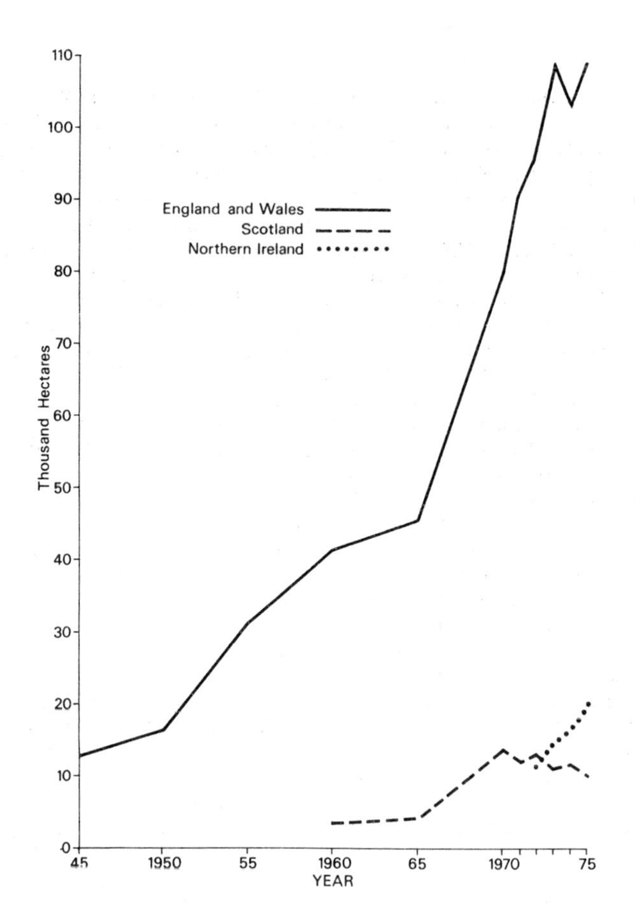

Figure 4.2. The growth in area underdrained in the United Kingdom 1945-1975.

greatly increased initially, but after two years net annual replenishment restarted and he estimated that the restoration of initial storage conditions would only take twelve years. Rycroft and Massey (1975) compared recorded hydrographs from the 170ha clay catchment at Shenley Brook End and simulated hydrographs which might have resulted if the basin had been mole drained. They concluded, from their simulation exercise and literature review, that

"1. There is no evidence to suggest that underdrainage increases flooding.

2. Mole drainage reduces the peak outflow rates from a catchment for heavy storms which are liable to give rise to flooding (because it inhibits surface runoff).

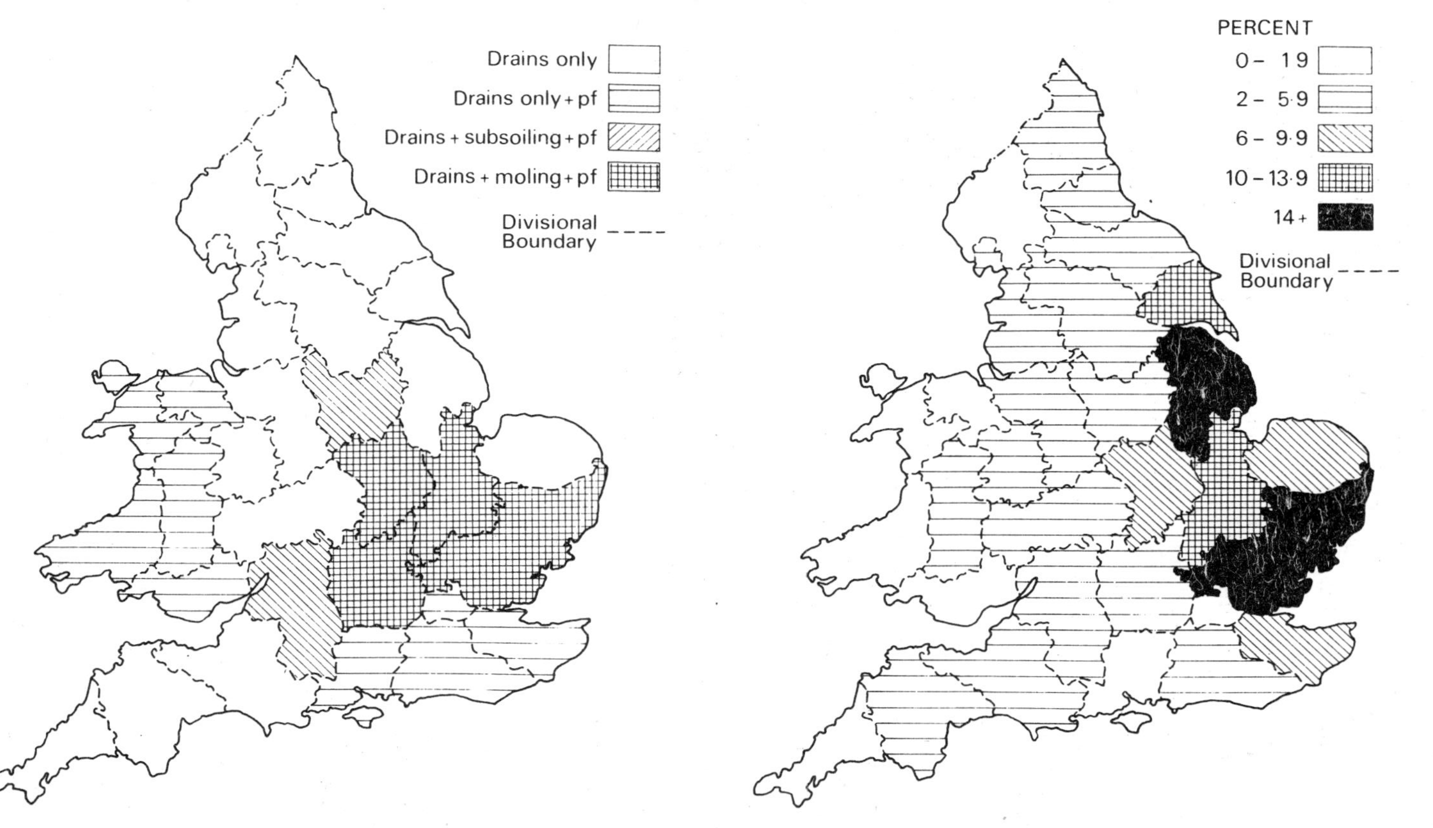

Figure 4.3. The spatial variation of underdrainage in the United Kingdom.

3. Mole drainage maintains watertable levels usually at 50cms depth, the watertable rises during storms storms but is quickly lowered after rainfall thus creating storage space for further rainfall.

4. Undrained clay catchments have limited storage which is filled up quickly during storms. Further rainfall then results in runoff.

5. An undrained waterlogged clay catchment remains wet for a considerable time after rainfall thus providing no buffer against further rainfall."

The inherent conflict between land drainage and nature conservation has been succinctly put by Humphries (1978) who said:

> "there is growing concern about increasing food production from a declining available acreage and about the incidence of higher flood levels. At the same time there is a growing awareness of the need to have regard for the preservation of the countryside and its ecological balance. ... the question (cannot) be resolved simply on economic grounds, it brings in broad social questions as well".

This concern for nature conservation is given statutory support by the Countryside Act 1968 and the Water Act 1973, both of which require Government and Water Authorities to have regard to the preservation and conservation of natural beauty. The growth in public concern is difficult to quantify but is reflected in the growth in membership of conservation-linked organisations detailed in Figure 4.4.

Figure 4.4. Membership of Conservation Bodies in the U.K.

ORGANISATION	1973	1974	1975	1976	1977
British Trust for Ornithology	6,935	7,038	7,281	7,100	7,348
Royal Society for the Protection of Birds	139,000	166,000	181,000	204,000	216,000
Society for the Promotion of Nature Conservation	84,412	97,180	100,660	111,134	115,211
Wildfowlers' Association of Great Britain and Ireland	28,000	29,000*	30,815	31,664	34,412
Wildfowl Trust	11,000*	12,000*	13,000*	14,000*	15,000

(* = estimated figures)

The problem is not as simple as a need to assess the impact of a new development and to make recommendations for management. Land drainage is an old established procedure and it is undoubtedly highly beneficial to agriculture. Many examples exist of valuable habitats created by drainage work eg. the Ouse washes in the Fenland and the Monmouthshire Levels (Scotter et al., 1977), and there is the problem of who should pay any costs attributable to conservation work. Moreover, land drainage engineers can be rather myopic. Miers (1974), for instance, writing about "design considerations" in a paper entitled 'The Civil Engineer and Field Drainage' does not mention or allude to nature conservation or ecological changes likely to result from drainage. However, elsewhere Miers (1975) said "outside of the fens and marshes the land drainage engineer has the greatest opportunity to conserve ... fauna and flora ... and improve the environment". How regrettable that he could not allude to this in his earlier discussion of drainage design. Cripps (1975) emphasised the point by saying "Miers' enlightened views contrast with some recent ... practises. Too much (land drainage) ... is still ... purely functional".

A large-scale example of the problem, but one which is typical of the hundred or thousands of small localised schemes, are the proposed drainage improvements in the Somerset Levels (Williams, 1977). This 687km^2 area of marshy farmland was drained piecemeal largely during the nineteenth century. Today, as a permanent pasture area, it is free of sea flooding and suffers only limited winter flooding. This habitat contains vast flocks of wintering wildfowl and waders on the soft pastures where insect food is plentiful. A high percentage of the British breeding population of certain birds and a magnificent flora add to the conservation interest. Finally, the thousands of miles of rhynes (ditches) provide a wide range of aquatic and semi-aquatic plants along with associated invertebrates. All of this is threatened by the desire of many farmers to increase productivity through a conversion to arable and improved grassland. In some areas this has begun privately and the agricultural advantages of the 1 metre lowering of the watertable are clear. Unfortunately, the ecologically detrimental effects of draining the wetlands and dessiccating the rhynes are also clearly apparent.

The Somerset Levels questions is far from being resolved but it is likely that worked out peat excavations will feature prominently in any agreed compromise. The proposal to form nine lakes from these pits could serve the needs of flood storage, water supply, land rehabilitation, recreation and nature conservation. In any event there is almost certain to be a public inquiry into the scheme. The conservation lobby scored a notable success when it forced a public inquiry into Southern Water Authority's scheme to drain the Amberley Wildbrooks. Moreover, the Minister ruled, after the inquiry, that the interests of conservation outweighed the potential agricultural benefit of that scheme.

Realisation of a need for a framework within which to resolve these conflicts emerged at the 1975 Conference on

Conservation and Land Drainage (Water Space Amenity Commission, 1975). Subsequently, Drummond (1977) argued that executing land drainage works and conserving our natural landscape is a matter of "management and compromise - accompanied by imagination and flair". The Working Party on Land Drainage and Conservation established by the Water Space Amenity Commission, published Draft Guidelines in 1978. They suggest "how, why and what action should be followed by those undertaking ... drainage ... to take account of ... nature conservation, landscape and amenity, fisheries and recreation". The Guidelines set out a system for consultation between engineers and conservationists and a set of practice notes for use by engineers during the design and execution of works. An example of active co-operation between land drainage engineers and conservationists is the work of Hollis and Kite (in press) on the Rivers Stort and Roding, N.E. of London. The environmental effects if conventional flood alleviation/river improvement schemes are being monitored, a model is being developed to forecast the ecological state of the rivers as it is recolonised and an ecological management plan is being prepared to maintain and enhance, where possible, the conservation interest of the rivers and their banks. There is, however, a long way to go before land drainage works are designed to satisfy hydraulic and ecological criteria. Cole (1976), Chief Engineer of the Ministry of Agriculture, Fisheries and Food, stated that "sheet steel piling has been increasingly used" for earth retaining structures, "it is just as important to remove weeds as ... any other constriction" and herbicides "have been very effective and, in general, have done little damage to the environment". George (1975) argued that a steel lined river gives "no chance of any natural vegetation ... at the river's edge; it is aesthetically unattractive and ... it is extremely dangerous" should people need to clamber from the stream. Even Johnson (1954), a predecessor of Cole at the Ministry of Agriculture, Fisheries and Food, stated that huge amounts of tree clearance had been undertaken "without a proper appreciation ... that ... tree growth, and the shade it gave, prevented or at least discouraged the growth of water weeds".

The future ecological state of the rivers of England and Wales is uncertain because the arguments for increased indigenous food production and the desire for flood protection are especially powerful. However, the conservation lobby is growing in number and in confidence and existing publications point to the minimal cost of, and in some cases, the financial savings from, a more "environmentally aware" style of river engineering. The increasing ability of the multi-functional Regional Water Authorities to utilize research results in their catchment control function (Addyman, 1979) suggests that, perhaps, a future management strategy will be the explicit inclusion of ecological yardsticks and landscape factors as well as hydrological criteria in the design of improved river channels.

NITRATES IN SURFACE WATER

The concentration of nitrate nitrogen in surface waters in some parts of the U.K. gives grounds for concern. The WHO recommended maximum for young infants of 11.3mg/1 and absolute maximum of 22.6mg/1 is often observed in rivers and the trend appears to be upward. There is similar concern for groundwater resources (eg. Foster and Crease, 1974) but this is discussed fully by Young and Gray, 1978; Young and Hall, 1977; and Young, 1979.

Scorer (1974) has shown that the average concentration of nitrate nitrogen in the Thames and Lee intakes of the Metropolitan Water Board rose from a long run average (1920-1940) of 2.7mg/1 and 4.0mg/1 to 12mg/1 and 17mg/1 respectively in the middle of 1974. He ascribed these changes primarily to unusually warm winters and low rainfall at the beginning of the 1970s but argued that the relative importance of growing discharges of nitrate rich sewage effluent, fertilizer applications, land drainage and increased cultivation of legumes was not yet known. The nitrate problem has continued in the Thames Water Authority area, especially in the River Lee. The Thames Water Authority Annual Report (1975) says the Lee "contained high concentrations of nitrate which on occasions were so high that abstraction for public supply had to be curtailed". Thames Water Authority (1976) said the concentration of nitrate in the River Lee at the Chingford intake has been above 11.3mg/1 for the first quarter of 1976. Thames Water Authority (1977) reported low concentrations of nitrate during the 1976 drought with levels rising rapidly sometime after flows had increased. Nitrate concentrations in excess of 11.3mg/1 persisted for several months in both the Thames and Lee. The Anglian Water Authority, which covers much of Eastern England, has also had severe problems. Anglian Water Authority (1977) said "no public supplies ... exceeded (22.6mg/1) during the year but severe problems were experienced with direct abstractions ... in Bedfordshire and Essex". Anglian Water Authority (1978) said relatively high peaks of nitrate concentrates were recorded in four cases. The use of a chalk source in West Norfolk was discontinued due to rapidly increasing nitrate levels". Greene (1978) stated that "there is a clear upward trend in many of the major rivers in Southern and Eastern England". Figure 4.5 shows the steady rise in the mean annual concentration of nitrate in the Great Ouse, the spectacular rise associated with the end of the 1976 drought and the high and rising concentration in the Mill River at the Bucklesham public water supply intake. This latter example is especially important because the small catchment has no significant volume of sewage effluent discharged into it. Slack (1977) writing about nitrate levels in rivers in Essex in south east England said "levels in excess of 20mg/1 of nitrate nitrogen have been recorded at times". It is important though, to note that the problem is not nationwide. The South West Water Authority (1977) said "rivers used for supply in this region normally contain very low concentrations of nitrate" and Northumbrian Water Authority (1978) do not discuss nitrate concentrations but include a table

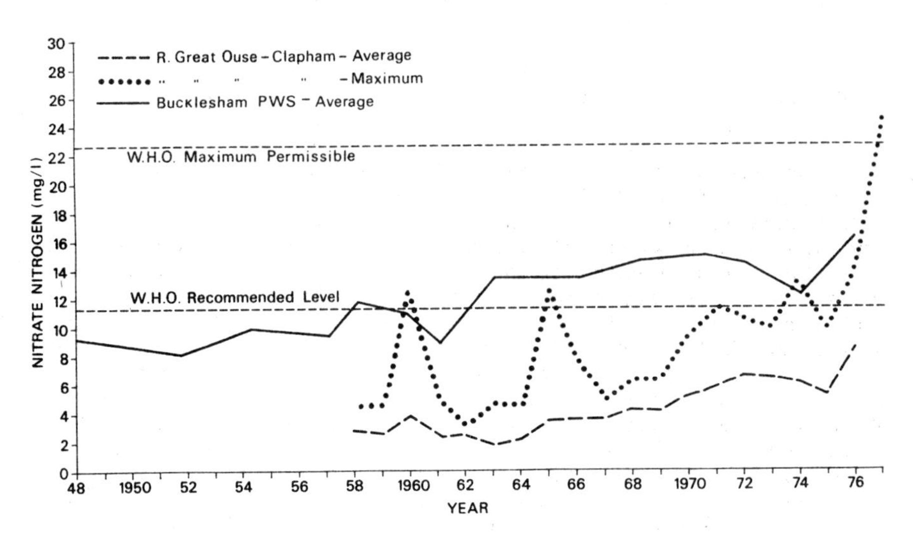

Figure 4.5. Nitrate Nitrogen Concentrations in some rivers in the East Anglian Water Authority area.

which shows that from 1972-77 the maximum nitrate nitrogen concentration recorded in their rivers was 4.79mg/1 with a normal average of between 1.0 and 2.0mg/1. What, then, is the cause of this disturbing rise in nitrate levels in the rivers of southern and eastern England in particular? Discussion of this question usually centres around fertilizers, sewage effluent, and natural processes being accelerated by unusual weather conditions.

At first sight the rise in the application rate of nitrogenous fertilizers, being contemporaneous with rising nitrate levels in rivers, is a powerful argument. Green (1973) has shown how the application rate of nitrogen rich fertilizer for crop and grass land in England has risen from around 4kg/ha in 1953 to 16kg/ha in 1972. He showed that the highest rate of application was in the south and east and in Cheshire. He showed that the greatest rate of increase in application of nitrogenous fertilizer is in the largely pastoral counties of the north and west, most counties exhibiting at least a five fold increase in application rates from 1957-72. Cooke (1976) reported that approximately 930,000 tonnes of nitrogen were added to the soil in the U.K. by fertilizers in 1973, whilst Holden (1976) in his study of Loch Leven found that nitrogen fertilizer usage in Kinross increased from under 300 tonnes/annum in 1952-54 to over 1000 tonnes/annum in 1970-72. This circumstantial evidence is not completely substantiated by small scale experimental studies nor is it proven con-conclusively in catchment scale studies.

Cooke (1976) reviewed three lysimeter experiments. 0.0004ha lysimeters were made at Rothamsted by enclosing undisturbed soil blocks in 1870. The soil has since remained free of crops and fertilizers, but it has been weeded regularly. From 1878-1905 the drainage contained 9.8mg/1 of nitrate nitrogen. In 1969 the hundredth year of the experiment, the drainage from April-November averaged 5mg/1 with 18mg/1 occuring in November. This experiment reveals that drainage from land which has never received fertilizer can contain substantial amounts of nitrates released or fixed by natural processes. Similar sized lysimeters near Aberdeen, established in 1914, were planted with grasses, root crops and cereals, and given modest amounts of nitrates in manure. It was found that drainage water from grassland carried little (0.5-3kg N/ha) nitrate whilst 4-11kg N/ha were lost from cropland. It is significant, though, that losses were almost as great from a lysimeter not given fertilizer. A recent lysimeter study at Jealott's Hill found that, without the use of N fertilizer, drainage from a grass sward contained .1mg/1 of nitrate nitrogen, that from growing clover, 29mg/1; that from bare soil after removal of clover, 34mg/1 and that from bare soil 42mg/1.

Cooke (1976) also reported studies on small (0.2ha) plots at or near Rothamsted all growing winter wheat. In the Broadbalk experiment, a plot with contemporary commercial rates of N fertilizer application had drainage water with 7mg/1 nitrate nitrogen, whilst unfertilized plots had 4mg/1 nitrate nitrogen in drainage. Water issuing from land drains at four agriculural sites in S.E. England has shown mean nitrate nitrogen concentrations of 11.7, 22.2, 20.0 and 9.8mg/1 with a range in one case of 11.5-36.5mg/1 and in another from 0.5-60.0mg/1. Johnson (1976) showed that long term use of N fertilizer does not produce a significant accumulation in the soil or sub-soil. He found that some crops recover 70-80 percent of the added N but the average uptake is only 46 percent. Hood (1976) studied drainage water from grassland that receives 250 and 750kg fertilizer N/ha/annum and is stocked heavily with dairy cows. He found that rainfall patterns affected nitrate losses but that overall less than 5 percent of the applied nitrate was leached, assuming no contribution from soil N reserves.

The evidence of these small scale studies is threefold. First, there can be substantial leaching of nitrates even when no fertilizers are used. Second, when fertilizers are used there is always incomplete uptake by the crop and therefore a high probability of leaching. Third, peak concentrations of nitrate in drainage water are related to the type of crop grown, the form of husbandry practised, the weather, the rate of fertilizer application and the timing of fertilizer application. Cooke (1976) confirms this relative position of fertilizer in his estimates of the N involved annually in U.K. farming systems (figure 4.6).

Catchment scale studies, comparing agricultural, industrial, sewage effluent and rainfall contributions to nitrate in rivers, nearly all point to farmland as the major source of nitrate nitrogen. Owens (1970) examined the nutrient budget of the River Great Ouse from the headwaters to Tempsford, 16kms below Bedford. He found that the 60 sewage effluent discharges contributed only 30 percent of the total flow of N. "It was therefore concluded that in the Great Ouse Basin, land drainage was a major source of nutrients". In reconnaissance studies of thirty three other U.K. rivers, Owens found that "the greater proportion of the nitrogen flow was derived from ... land drainage". He was, however, unable to demonstrate with his rather poor data, a clear relationship between

	Thousands of tonnes
Soil N (natural fixation, mineralization)	1,300
Farmyard manure	300
Fertilizer	900
Excreta dropped on grass	500
Total	3,000

Figure 4.6. Estimates of the N involved annually in U.K. farming systems. (Source: Cooke, 1976)

nitrogen concentraion in river water and the application rate of nitrogenous fertilizer. Moreover, using data for three Essex rivers and six Yorkshire rivers, he found that "although the amounts (of N fertilizer) applied per unit of catchment have increased substantially, no marked increases in the loads carried by the river at any given flow are apparent". Slack (1977) charted the seasonal cycle of nitrate in the Chelmer, Blackwater and Stour in Essex from 1970-74. High winter concentrations were associated with high flows and "as the contribution from sewage effluent (is) fairly constant, the increased nitrate can only be derived from drainage on arable land which forms the majority of the catchment area of the rivers". Greene (1978) similarly found "high nitrate concentrations are invariably associated with high flow conditions in all ... rivers in Anglia signifying that the major nitrate load is associated with land drainage". He said that "current information appears to indicate that (recent) increases (in nitrate nitrogen in river water) are primarily associated with the increasing intensity of and/ or improvement in arable farming in the region in the last twenty years". The nitrate nitrogen content of rainfall is modest. Mean concentrations of 1.09 (6.8 kg/ha/annum), 0.93 (5.1 kg/ha/annum) and 1.17mg/1 (6.7 kg/ha/annum) for sites in south east England have been reported by Williams (1976)for the period 1969-73. Troake and Walling (1975) measured a concentration of 0.25mg/1 (2.75 kg/ha/annum) for a maritime location in S.W.England and 0.1-4.5mg/1 has been quoted as the range for the U.K. (Greene, 1978).

The nitrate inputs to Loch Leven have been monitored by Holden (1976). He found the influent streams to be the major source of nitrogen. Their agricultural catchments produced nitrogen yields ten times those of unculivated moorland streams. He estimated that the annual loss of nitrate of 33kg/ha was possibly more than one third of the nitrate applied through fertilizer. Troake et al. (1976) found that for the period 1971-74 the N load of two experimental catchments which are free of sewage effluent discharges was 50 percent and 26 percent of fertilizer application; the former figure deriving from a catchment with particularly short steep slopes which would favour transmission of soil water to the stream as throughflow. An analysis of the nutrient budget of Alderfen Broad in East Anglia (Phillips, 1977) revealed that "input of nitrate nitrogen via the inflow was the most important source of nitrogen ... Maximum nitrate nitrogen input occurred during November due to a very high concentration ... (14mg/l) combined with a high flow rate".

The evidence, therefore, points clearly to farmland as the major source of nitrate in Britain's rivers but the evidence does not pinpoint fertilizers as the major source. The efficiency of natural nitrate production by mineralization and atmospheric fixation and the rapidity of rainwater leaching were dramatically demonstrated by studies of the effects of the 1976 drought and the subsequent wet months. The period May 1975 - August 1976 with 757mm of rainfall over England and Wales was the driest 16 month period since records began in 1727 and has a recurrence interval of more than 1000 years. Many riverflows were lower than ever recorded before, often for 6 months or more. The estimated recharge for many U.K. aquifers has an estimated return period of 1 in 100 years (Central Water Planning Unit, 1976).

Writing about the Anglian Water Authority area, Davies (1978) said "the most notable change in the quality of river water during the (post-drought) period was the rapid increase in the concentration of nitrates ... in surface waters ... one river reached 40mg/l nitrate nitrogen, while concentrations with high flows gave an extremely large total discharge. Davies argued that the shortage of summer rain limited nitrate uptake by crops, in fertilizer usage and rising nitrate levels in streams. Moreover, underdrainage has been most extensive in the south and east where there is the heaviest rate of application of N fertilizer. There is evidence to support this view from soil science studies. Waters (1977) has shown how soil temperature is higher in spring on drained land than on undrained land. She showed, by means of a derived relationship between bacterial activity and temperature, that underdrainage might increase peak bacterial activity by 1.5 times. Similarly, Young and Hall (1977) concluded from a literature review that a "pronounced release of nitrogen from soil occurs with the ... improvement in the drainage of wet soils". They also found that "retention of nitrogen in soil is promoted by the absence of aeration". A final piece of circumstantial evidence is the incidence of very high nitrate concentrations, quoted above, for tile drain outlets.

There is a dearth of published studies on the nutrient effects of underdrainage. Work, though, is underway ointly between the Institute of Hydrology and the Ministry of Agriculture, Fisheries and Food (Ins. of Hydrology, 1987) with a lysimeter experiment at Plynlimon, a catchment scale study of water quality and fertilizer use at Shenley Brook End in Central England and a study of drained and undrained plots near Grendon Underwood.

CONCLUSION

Agricultural practices do, clearly, modify the hydrological regime in the United Kingdom. The high rate of evaporation observed from water held in interception storage in trees explains why afforestation reduces water yield in upland areas where rainfall is frequent and plentiful. Given the economic and silvicultural need to grow conifers there is little that can be done to mitigate this loss in yield other than to take note of it for the design of water resource systems. Land drainage does not increase the magnitude of flood flows. Indeed, the presence of soil storage can reduce the size of moderate floods. Field underdrainage can be very successful in reducing the water table in many soils but its beneficial effects are not large everywhere. The existence of extensive underdrainage appears to marginally increase low flows in streams. Land drainage and river improvement works reduce, and may devastate, the nature conservation interest of an area. There is a need for close consultation and liaison between engineers and ecologists. The establishment of ecological and hydraulic criteria for river improvement works would be useful.

Increasing concentrations of nitrate nitrogen in surface waters is one of the most pressing problems facing the water industry in the U.K. Heavy or ill-timed dressings of fertilizer can produce high rates of nitrate leaching. There is no conclusive proof that the ever increasing use of nitrogenous fertilizer is directly responsible for the increase in nitrate observed in streams, but all studies of fertilizer use show an incomplete uptake by the crop. Investigations of unfertilized soils do show substantial yields of nitrate in drainage water and the wet months after the 1976 drought attested to the efficiency of natural fixation and mineralization. There is circumstantial and scientific evidence to suggest that field underdrainage is a significant factor in increasing the movement of soil nitrates into streams, but a dearth of research on this subject prevents definitive conclusions.

REFERENCES

1. Addyman, O.T., 1979, The use of research in catchment control, in: *Man's Impact on the Hydrological Cycle in the U.K.*, ed. Hollis, G.E., 234-250. (Geo Books Norwich)

2. Anglian Water Authority, 1977, *Annual Report and Accounts,* 1976-77.

3. Anglian Water Authority, 1978, *Annual Report and Accounts,* 1977-78.

4. Armstrong, A.C., 1977, An analysis of crop yield and other data from the Drayton experiment, *Field Drainage Experimental Unit, Technical Bulletin,77/3.*

5. Binns, W.O., 1979, The hydrological impact of afforestation in Great Britain, in *Man's Impact on the Hydrological Cycle in the U.K.,*ed. Hollis, G.E., 55-70. (Geo Books Norwich)

6. Calder, I.R., 1976, The measurement of water losses from a forested area using a "natural" lysimeter, *Journal of Hydrology, 30,* 311-325.

7. Calder, I.R., 1977, A model of transpiration and interception loss from a spruce forest in Plynlimon, Central Wales, *Journal of Hydrology, 33,* 247-265.

8. Calder, I.R., 1979, "Do trees use more water than grass?", *Water Services, 83 (995),* 11-14.

9. Central Water Planning Unit, 1976, The 1975-76 Drought: A Hydrological Review., *Central Water Planning Unit, Technical Note 17.*

10. Clarke, R.T., and McCulloch, J.S.G., 1979, The effect of landuse on the hydrology of small upland catchments, in: *Man's Impact on the Hydrological Cycle in the U.K.,*ed. Hollis, G.E., 71-78.(Geo Books Norwich)

11. Clarke, R.T., and Newson, M.D., 1978, Some detailed water balance studies of research catchments, in: *Scientific Aspects of the 1975-76 Drought in England and Wales, (Royal Society, London)* 21-42.

12. Cole, G., 1976, Land drainage in England and Wales, *Journal of the Institution of Water Engineers and Scientists, 30,* 345-367.

13. Cooke, G.W., 1976, A review of the effects of agriculture on the chemical composition and quality of surface and underground waters, in: *Agriculture and Water Quality, Ministry of Agriculture Fisheries and Food Technical Bulletin, 32,* 5-57.

14. Cripps, J., 1975, Contribution to the discussion, in: *Proceedings of the Conservation and Land Drainage Conference, Water Space Amenity Commission,* 45-47.

15. Davies, A.W., 1978, Pollution problems arising from the 1975-76 drought, in: *Scientific Aspects of the 1975-76 Drought in England and Wales, (Royal Society, London)* 97-107.

16. Drummond, I., 1977, Conservation and Land Drainage, *Water Space, 11,* 23-30.

17. Foster, S.S.D., and Grease, R.I., 1974, Nitrate pollution of chalk groundwater in East Yorkshire, *Journal of the Institution of Water Engineers and Scientists, 28,*178-194.

18. Law, F., 1956, The effect of afforestation upon the yield of water catchment areas, *Journal of the British Waterworks Association, 38,* 489-494.

19. May, J., and Trafford, B.D., 1977, The analysis of the hydrological data from a drainage experiment on clay land, *Field Drainage Experimental Unit Technical Bulletin 77/1.*

20. Miers, R.H., 1974, The civil engineer and field drainage, *Journal of the Institution of Water Engineers and Scientists, 28,* 211-223.

21. Miers, R.H., 1975, A guide for land drainage engineers on conservation, amenity and recreation, in: *Proceedings of the Conservation and Land Drainage Conference, Water Space Amenity Commission,* 43-45.

22. Northumbrian Water Authority, 1978, *Fourth Annual Report and Accounts, 1977-78.*

23. Owens, M., 1970, Nutrient balances of rivers,*Water Treatment and Examination, 19,* 239-247.

24. Painter, R.B., Blyth, K., Mosedale, J.C., and Kelly, M., 1974, The effect of afforestation on erosion processes and sediment yield, in: *Effects of man on the interface of the hydrological cycle with the physical environment, International Association of Hydrological Scientists Publication 113,* 62-68.

25. Pereira, H.C., 1976, Final Discussion: Summary and Conclusions, in: *Agriculture and Water Quality, Ministry of Agriculture, Fisheries and Food, Technical Bulletin 32*, 467-469.

26. Phillips, G.L., 1977, The mineral and nutrient levels in three Norfolk Broads differing in trophic status, and an annual mineral content budget for one of them, *Journal of Ecology, 65*, 447-474.

27. Rutter, A.J., 1963, Studies in the water relations of Pinus sylvestris in plantation conditions. I Measurement of rainfall and interception, *Journal of Ecology, 51*, 191-203.

28. Rutter, A.J., Kershaw, K.A., Robins, P.C., and Morton, A.J., 1971, A predictive model of rainfall interception in forests. I Derivation of the model from observations in a plantation of Corsican pine, *Agricultural Meteorology, 9*, 367-384.

29. Rutter, A.J., Morton, A.J., and Robins, P.C., 1975, A predictive model of rainfall interception in forests. II Generalization of the model and comparison with observations in some coniferous and hardwood stands, *Journal of Applied Ecology, 12*, 367-380.

30. Rutter, A.J., and Morton, A.J., 1977, A predictive model of rainfall interception in forests. III Sensitivity of the model to stand parameters and meteorological variables, *Journal of Applied Ecology, 14*, 567-588.

31. Rycroft, D.W., and Massey, W., 1975, The effect of field drainage on river flow, *Field Drainage Experimental Unit Technical Bulletin, 75/9*.

32. Scorer, R., 1974, Nitrogen: A problem of decreasing dilution, *New Scientist, 62(895)*, 182-184.

33. Scotter, C.N.G., Wade, P.M., et al., 1977, The Monmouthshire Level's drainage system: Its ecology and relation to agriculture, *Journal of Environmental Management, 5*, 75-85.

34. Slack, J.G., 1977, Nitrate levels in Essex river waters, *Journal of the Institution of Water Engineers and Scientists, 31*, 43-51.

35. South-West Water Authority, 1977, *Annual Report and Accounts 1976-77*.

36. Thames Water, 1975, *Annual Report and Accounts 1974-75*.

37. Thames Water, 1976, *Annual Report and Accounts 1975-76.*

38. Thames Water, 1977, *Annual Report and Accounts 1976-77.*

39. Troake, R.P., Troake, L.E., and Walling, D.E., 1976, Nitrate loads of South Devon streams, in: *Agriculture and Water Quality, Ministry of Agriculture, Fisheries and Food Technical Bulletin 32,* 340-351.

40. Troake, R.P., and Walling, D.E., 1975, Some observations on stream nitrate levels and fertilizer application at Slapton, S.Devon, *Rep. Trans. Devonshire Association Advancement of Science, 107,* 77-90.

41. Walling, D.E., and Foster, I.D.L., 1978, The 1976 drought and nitrate levels in the R.Exe basin, *Journal of the Institution of Water Engineers and Scientists, 32,* 341-352.

42. Water Data Unit, 1975, Water Data 1974, Water Data Unit, Reading.

43. Water Data Unit, 1977, Water Data 1975, Water Data Unit, Reading.

44. Water Space Amenity Commission, 1975, *Proceedings of the Conservation and Land Drainage Conference.*

45. Water Space Amenity Commission, 1978, *Annual Report 1977-78.*

46. Water Space Amenity Commission, 1978, *Conservation and land drainage guidelines: Draft for consultation.*

47. Waters, P., 1977, The effect of drainage on soil temperature at Drayton EHF, *Field Drainage Experimental Unit, Technical Bulletin 77/6.*

48. Wilcock, D.N., 1979, The hydrology of a peatland catchment in N.Ireland following channel clearance and land drainage, in: *Man's Impact on the Hydrological Cycle in the U.K.,* ed. Hollis, G.E. 93-108.

49. Williams, R.J.B., 1976, The chemical composition of rain, land drainage and borehole water from Rothamsted, Broom's Barn, Saxmundham and Woburn Experimental Stations, in: *Agriculture and Water Quality, Ministry of Agriculture, Fisheries and Food Technical Bulletin 32,* 174-200.

50. Williams, R., 1977, The Somerset Levels: A case for conservation? *Water Space, 12,* 15-18.

51. Young, C.P., and Gray, E.M., 1978, Nitrate in Groundwater, *Water Research Centre Technical Report 69.*

52. Young, C.P., and Hall, E.S., 1977, Investigations into factors affecting the nitrate content of groundwater, in: *Groundwater Quality - Measurement, Prediction and Protection,* Water Research Centre, Medmenham, U.K., 443-469.

53. Young, C.P., 1979, The impact of agricultural practices on the nitrate content of groundwater in the principal U.K. aquifers, *Proceedings of the International Conference on Environmental Management of Agricultural Watersheds,* Smolenice, Czechoslovakia.

5. Granite structures and landforms

A.J. Gerrard

Introduction

The influence of rock types and structures on the evolution of landforms has long been recognised. Traditional treatments of rock or structural control have often remained at a very simplistic level with few attempts to study, in detail, the properties of rocks. Yatsu (1966) was determined to discard the vague and sometimes ambiguous generalisations that can be found in the literature and to reveal the detailed and specific relationships involved. Rock control theory is concerned with the influences of rock properties on the formation of landforms. But the phrase rock control should not be seen as a 'magic cloak' to avoid clarifying the specific mechanisms involved. Conclusions obtained from juxtaposition of landforms and geology are merely prospective hypotheses. The main problem is to examine these hypotheses and to resolve the questions of the mechanisms and processes of rock control.

The strength of a rock will differ according to the geomorphological processes involved. Even if the processes and rock types are the same, the rock strength will not be the same if the rate of operation of the processes differs. Thus, sound granite is very resistant to simple mechanical abrasion but is susceptible to physical breakdown and chemical decay and there is a great contrast in erosional mobility between unaltered granite and its weathered products. Rock strength will also depend on the type of destruction brought about by specific processes. Running water breaks down rock by wear and impaction failure, whereas glacial action involves compression failure as well as wear. The action of sea waves involves, wear, impaction failure and both compression and tension failure. Thus, both compressive and tensile strengths of rocks are important.

Yatsu (1966), in his conclusion, states that his aim was to emphasise the importance of physico-chemical and mechanical understanding of rocks in geomorphological studies and to explain such thinking and methods of study.

It is the aim of this essay to examine some of the characteristics of granite in this way and to suggest avenues for further investigation. It is important to review the possible modes of formation of the major structural features of granite in order to assess the relationships between these features and landforms. The nature of these structural features, such as joint orientation, spacing and nature of joint surfaces, determines the specific response of weathering and erosional processes. These factors also determine the strength of the rock and are examined in some detail. Most of the specific examples are taken from the granite masses of south west England and attention will be focussed on the structural aspects, such as jointing, and not on the mineralogical variations in the granite.

TYPES OF GRANITE STRUCTURES

The importance of joints, as influencing factors in the development of granite landforms, has been stressed by most workers, and a good review of much of this work has been provided by Thomas (1974). Although joints are the most conspicuous structural features, other features, such as rift, grain, micro-jointing and micro-cracks, have a significant influence on rock behaviour. The relationship between landforms and jointing is also complicated by the fact that some joints are of primary origin whereas others are secondary in nature and it is therefore essential to be able to distinguish between the two types.

Sheeting joints

Granite is characterised by curved sheet joints which match approximately the surface contours of the landscape. Some of these joints may be primary structural features or L-joints in the classification of Cloos (1936). Tensional strains are set up during emplacement and cooling which lead to primary sheeting structures (Oxaal, 1916; Meunier, 1961). Alternatively, sheeting may be created by the process of unloading on release of primary confining pressures. Unloading was first suggested by Gilbert (1904) and since then has been supported by many workers (e.g. Jahns, 1943; Kieslinger, 1958; Bradley, 1963). The mechanics of the process have been described by Brunuer and Scheidegger (1973). Unloading has been accepted by most workers in glaciated areas (e.g. Lewis, 1954; Harland, 1957), where it is caused by the loading of valley floors by ice, followed by rapid unloading when the ice retreats. Addison (1981), however, has warned against a universal acceptance of the process based on inadequate data.

Soen (1965) has suggested a slightly different mechanism for sheeted granite in Greenland. These granites are associated with a negative gravity anomaly and, because of a mass deficiency in the crust, gravitational forces tend to raise the deficient rock masses above their surroundings. Initially, a vertical compression is exerted on the higher levels of the granite but, once the mass deficiency is compensated, uplift ceases and a relative decompression takes place, favouring large scale sheeting in the near surface rocks.

The unloading process, as an hypothesis, is difficult to test in the field because the same evidence can be used to argue for both secondary unloading and for a primary origin. Selby (1971) has examined the unloading hypothesis in Antarctica and in the Namib Desert of Southern Africa. In Antarctica, sheeting joints follow the changes and breaks of slope almost exactly. Rectilinear slopes increase in length by elimination of the free face above and the curved jointing is interpreted as a stress-release phenomenon adjusting to the evolving shape of the concavity. In the Namib Desert, multiple-domed forms are explicable by the incision of channels along major vertical joints with the development of sheeting parallel to the slopes above these channels. A primary origin would imply the existence of multi-domed intrusion forms.

Differences of opinion still exist concerning the origin of the sheeting or pseudo-bedding planes in the granite of south west England. Brammall (1926) and Edmonds et al. (1968) have argued in favour of the joints being original structural features whilst Waters (1954) believes they have developed in response to the evolving topography.

Many workers now accept that the majority, if not all, of the sheeting is the result of the unloading mechanism, although a number of problems have been raised by Twidale (1972, 1973). If an extensive network of open vertical joints existed prior to unloading, it would be expected that compressive stresses would be released along these joints reducing the necessity for stresses to be released in sheeting joints. Thus, the distinct possibility exists, following the suggestion of Chapman (1958), that many of the vertical or steeply dipping joints now prominent on granite landforms are relatively late features, probably opened up after the sheeting developed. This idea is examined later.

A-tents

This is the name given by Jennings and Twidale (1971) to raised and fractured slabs delimited by a curvilinear joint plane parallel with the rock surfaces. The slabs are often raised 50-80 cms above the rock surface and are usually fractured in the centre of the arch. Field evidence suggests that these features are the result of compression in the rock. Buckling on thin sheets caused by the expansive strain in the rocks was also noted by Dale (1923).

Laminae

Many granite blocks possess flakey, flaggy or plate-like skins of rock which have formed parallel to the rock surface. These plates may be up to 40 cms thick but are more commonly in the range 1-5 cms. They conform closely to the rock surface and overhang rounded boulders. Selby (1977), in examining these features in the Namib Desert, found little evidence of physical or chemical weathering and argued that the lamination was also a stress release or unloading phenomenon.

Strain energy can be stored in any granular material (Emery, 1964). Some of this primary strain energy is released through jointing, as described earlier, but, even in closely jointed rock, some residual strain energy remains. Different materials will possess different mechanical properties and when load is released, some of the stored strain energy will also be released. Weathering is one of the processes which leads to a release of some of this strain energy and the ease with which weathering takes place may be dependent on the previous loading history of the rock. Durrance (1969) has argued that Dartmoor tors represent zones of low initial strain energy level. Little residual strain energy remains after jointing to aid intergranular breakdown. Areas of growan would correspond to zones with high initial strain energy levels and the later development of a tight joint system. A high residual strain energy would remain and aid the further breakdown of the internal rock structure.

Vertical or steeply dipping joints

Joints begin to form in a rock mass as soon as it is capable of brittle fracture and it is now believed that the fundamental fracture pattern is established early in a rock's history. In rock deformation studies, the term extension fracture has been applied to fractures that form normal to a tensile least principal stress (Secor, 1965). Shear fractures are inclined to the directions of principal stress and can occur in two conjugate sets.

The most usual geological classification of this type of jointing follows that of Cloos (1936). Joints arranged at right angles to the original flow lines (Q-joints) are tensional open joints liable to be infilled with more fluid magma residues. Joints parallel to the original flow lines (S-joints) are best developed near the roof of the granite intrusions where flow lines approach the horizontal. Crosby (1893) has suggested that compressive strains may have been supplemented by vibrating strains in causing joints and Becker (1905) has shown that four or even more systems of joints may be due to a single force. Diagonal joints, usually at 45 degrees to the pressure directions, are not uncommon and folds or quasi-anticlinals created by tensional forces directed away from the strike of the Q-joints can also occur. Where faulting has occurred, diagonal joints or Riedel shears may be created and it has been suggested that some of the jointing in north-east Dartmoor is of this type (Blyth, 1962).

Analysis of fresh joint surfaces reveals faint plumose markings which become coarser near the fringe of the joint plane. These were first described by Woodworth (1896) but subsequently many other workers have described similar structures. Hodgson (1961) and Roberts (1961) have argued that their presence indicates that there has been no lateral movement parallel to the fracture surface. Roberts (1961) also suggested that the markings originate from a rapid opening of the joints in a direction normal to the fracture plane.

These markings and other characteristics indicate that joints are tensional features but theoretical analyses show that absolute tension is impossible below very shallow depths. However, this can be reconciled with the theory developed by Griffith (1921), which is based on the assumption that solids are filled with minute cracks. Under differential pressures tensile stress will occur near the ends of cracks even when all the principal stresses are compressive. Thus, failure occurs by the growth of such cracks in response to the tension.

Headings

This is the term used by Dale (1923) to describe situations where joints occur so close together as to break up the rock into small blocks. Dale (1923) has described one quarry in Massachusetts where joints in headings were only 2-5 cms apart. Headings allow easy access for surface water and the granite is usually partially decomposed. In some of the deeper quarries in Massachusetts, a heading at the surface would disappear at depth or a heading would appear abruptly 30 metres or so below the surface. The character of the joints indicates that complex stresses have been involved in their formation.

Subjoints or microfractures

Subjoints are minute fractures which branch off the main joints and which penetrate the rock for a few centimetres. Many of these subjoints are filled with secondary quartz, feldspar or biotite and have been described by Woodworth (1896) as joint fringe and feather fractures. They will allow water, circulating in the main joints, to penetrate the rock mass for a short distance, causing preferential weathering and considerably weakening what was originally sound, resistant rock.

Microcracks

Granite and many other igneous rocks possess numerous microcracks which may occur completely in one grain, cross or follow grain boundaries. Irfan and Dearman (1978) make the important distinction between structural microcracks, which are present in unweathered rocks and are generally straight and tight, and weathering microcracks, which are wider and commonly sinuous.

As the rock becomes more weathered, grain boundaries become stained and opened and new microcracks are formed. Seven types of microcracks formed in this way in granite have been listed by Irfan and Dearman (1978). These are stained grain boundaries, open grain boundaries, stained microcracks in quartz and feldspars, infilled microcracks in quartz and feldspars, clean, transgranular microcracks, filled or partially infilled microcracks and pores in plagioclase feldspars.

Microcracks have a major influence on rock quality and strength, but are extremely difficult to quantify. Dixon (1969) quantified microcracks by counting the number of fractures intersected when making a squared traverse, 10 mm on each side across a thin slice. He established linear relationships between total microcrack intensity and unconfined compressive strength and between permanent strain and unfilled microfracture intensity. Attempts have also been made to relate microcracks to certain elastic properties of the rock (e.g. McWilliams, 1966; Willard and McWilliams, 1969). Onadera et al. (1974), using the number and width of microcracks as an index of physical weathering, found a linear relationship between effective porosity and density of microcracks. They also found that the mechanical strength of the granite decreased rapidly as the density of cracks increased. These results were echoed in the conclusions of Simmons et al. (1975), which emphasised that values are needed for parameters such as crack dimensions, density, distribution and grain.

Rift and grain

The rift in granite is an obscure microscopic foliation along which the rock splits more easily than in any other direction. The grain is a foliation at right angles to the rift along which the rock splits with an ease second only to that along the rift. An admirable summary of these phenomena and a review of the early literature is provided by Dale (1923). Dana (1876) argued that granite often has a direction of easiest fracture due to the fact that the feldspar crystals have approximately a uniform position in the rock bringing the cleavage planes into parallelism. This does occur in some porphyritic granites, but, in the majority of cases, feldspar cleavages intersect the rift face at all angles. Other early suggestions invoked compressive strains as the cause. Zirkel (1894) argued that rift might be the result of strain caused by pressure from one side only which was not released by jointing and Suess (1913) suggested that horizontal rift or grain was the primary effect of compressive strain and that vertical rift and grain were secondary effects of the same strain.

The earliest detailed descriptions of rift and grain were by Tarr (1891) and Whittle (1900). The descriptions by Tarr seem to indicate microscopic cracks, most of which meander across both feldspar and quartz crystals. One of the sketches by Whittle shows quartz and feldspar grains crossed by a parallel set of lines corresponding to the rift planes. The studies by Dale (1923) revealed fluid inclusions in the quartz grains corresponding approximately in direction with that of the rift and grain.

It is possible to make several generalisations based on this previous work. Rift and grain structure appears to consist of minute cracks, 0.1 to 1.3 millimetres apart, crossing quartz particles and extending into the feldspar crystals. Both rift and grain are independent of flow structure and also of sheet structure. The rift is often parallel to mica plates and may be parallel to the long axes of feldspar phenocrysts. Rift and grain fractures are parallel to, or coincide with, sheets of fluidal cavities in the quartz crystals and secondary minerals often form within the cracks.

Experiments conducted by Douglass and Voight (1969) have shown that rift and grain planes are aligned parallel to microfracture concentrations and are planes of minimal compressive and tensile strength. This is consistent with theoretical and empirical work on the orientation of critical cracks in a tensile stress field. Cracks contribute to both non-linear stress-strain behaviour and rock anisotropy. The microfracture preferred orientation planes also represent planes of reduced rock cohesion and could be effective as stress-concentration elements.

NATURE OF GRANITE JOINTS

Joint orientation

Analysis of joint orientation data often indicates that one or two major preferred directions are consistently present within individual intrusive masses. Mean tension Q-joints strike approximately 90° on Dartmoor (Gerrard, 1974), 73° on Bodmin Moor (Exley, 1965), 67° in St. Austell Moor (Exley, 1958) and 53° on Carmenellis (Austin, 1960). Mean compression S-joints strike 180° on Dartmoor, 171° on Bodmin Moor, 168° on St. Austell Moor and 155° on Carmenellis. Preferred orientations impart a lineament to the landscape which is reflected by ridges and valleys. On Dartmoor, the differentiation of the landscape into positive and negative areas (Waters, 1957) and drainage networks with north-south and east-west alignments reflect this control. This close relationship is well seen in east Dartmoor, around Hound Tor (figure 5.1). But, a close inspection of the joint orientation diagrams indicates that steeply dipping joints are not confined to these preferred directions (figure 5.2). Some tors, such as Hound Tor, are dominated by two orthogonal sets of joints, but others, such as Bonehill Rocks, exhibit a greater variability.

This variability has led some workers to make a distinction between these joints. Thorp (1967) has differentiated

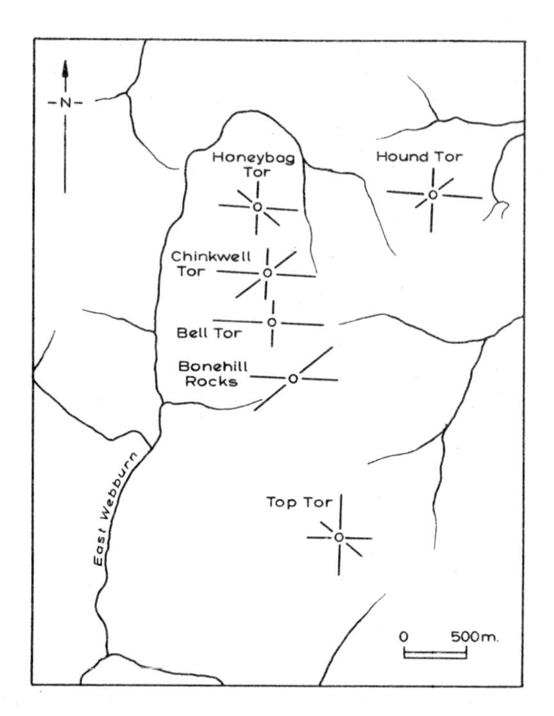

Figure 5.1. Orientation of vertical joints exposed on tors in east Dartmoor.

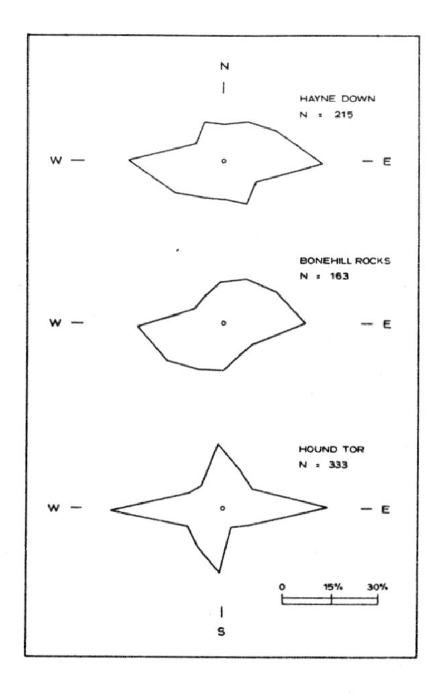

Figure 5.2. Variability in vertical joint orientation for three east Dartmoor tors.

between major or master joints and minor joints in the Nigerian granite masses. Master joints extend over considerable distances with an extent greater than the individual valleys developed along them. These joints are often gently curved in plan. Minor joints vary considerably in length, frequency and orientation and are rarely more than two or three kilometres in length. The frequency of jointing was often related to rock texture, with the closer network occurring on the finer-grained rocks.

The direction of the vertical joints is often so closely related to the topography, i.e. parallel and at right angles to the contours, as to suggest that they have also formed during the process of unloading. This possibility was suggested by Chapman (1958) and certainly seems likely on many parts of Dartmoor (Gerrard, 1974). The available evidence indicates that granite initially possesses many directions of potential jointing but only a few, in any small area, ever develop into true joints. Such a compromise view has also been suggested by Exley (1958) for sheeting in the St. Austell granite mass of Cornwall. The dominant or master joints have formed from the sets most strongly impressed upon the rock at the time of its formation. These potential joints could have been so impressed on the rock that they opened, in preference to weaker sets, even where conditions were less favourable for expansion. Underground observations have shown that joints open up and become loosened wherever openings or tunnels exist (Heitfeld, 1966). This means that joints now visible on rock outcrops have opened up at different times which makes it extremely difficult to decide whether the landscape has evolved in response to the jointing or vice versa. In most cases it is a combination of the two possibilities but great care is required in establishing the specific relationship.

Joint spacing

The spacing of the joints is probably the most important factor in defining both the gross shape and fine detail of granite landforms. But joint spacing is not easy to measure and consequently several different methods have been proposed. Some workers have suggested types of area sampling. Thus, Da Silvera et al. (1966) advocated that sampling should be based on portions of the rock surface areas to form a mesh. The other main method is to use a line sample and record every joint that intersects the tape or line. The joint spacings reported here for the granites of south-west England have been measured on two mutually perpendicular faces, using the line sample method.

Subjective descriptive terms have been suggested to cover the complete range of joint spacings likely to be encountered in any rock type (figure 5.3). This may need to be modified for particular rock types since the majority of the joint spacings on exposed granite faces in south-west England fall in the wide and extremely wide categories.

Description	Spacing
Extremely wide	>2 m
Very wide	600 mm - 2 m
Wide	200 - 600 mm
Moderately wide	60 - 200 mm
Moderately narrow	20 - 60 mm
Narrow	6 - 20 mm
Very narrow	<6 mm

Figure 5.3. Suggested classification for joint spacing (after Anonymous, 1977).

The spacing or frequency of the joints, so crucial to most theories of tor evolution, will also be influenced by the unloading process. Tors, in the theories of Linton (1955) and Palmer and Neilson (1962), are supposed to represent residuals isolated because of wider spaced vertical jointing. Detailed measurements of vertical joint spacing demonstrates that a great variability exists both within and between tors (figure 5.4). Most of the tors possess very closely spaced joints and it is difficult to envisage even more closely spaced jointing occurring in the gaps between tors. The same is true of the spacing of the horizontal joints, although there is less variability (figure 5.5). A combination of these data allows typical block sizes to be constructed for each tor which stresses the gross differences that exist between tors (figure 5.6).

It is difficult to imagine tors remaining as features if the joints had been always as open and as frequent as they are now and it is distinctly possible that the present joints are relatively late features. Observations in tunnels and mines have shown that the spacing of joints increases and the size of the openings decreases with depth (Leeman, 1958). In a series of observations in bore holes, Snow (1968) has shown that all joint spacings, in a variety of rocks, exceeded 4 metres at a depth of 90 metres. Also, the size of the openings ranged from 75 to 400 microns in the upper 9 metres to between 50 and 100 microns from 15 to 60 metres deep. This has great significance when considering the evolution of tors.

The spacing of the joints is also related to the type of granite. The joint spacings, noted so far for south-west England, are for the coarse-grained or megacrystic granite. Sill-like intrusions of fine-grained microgranite also occur which possess much closer spaced joints. The contrast, when exposed on tors, can be striking (figure 5.7). In the case of Bench Tor, the fine grained granite has been preferentially attacked by weathering and various erosional processes, producing a marked 'rock shelter' (figure 5.8). This is partly due to the closer spacing of the joints.

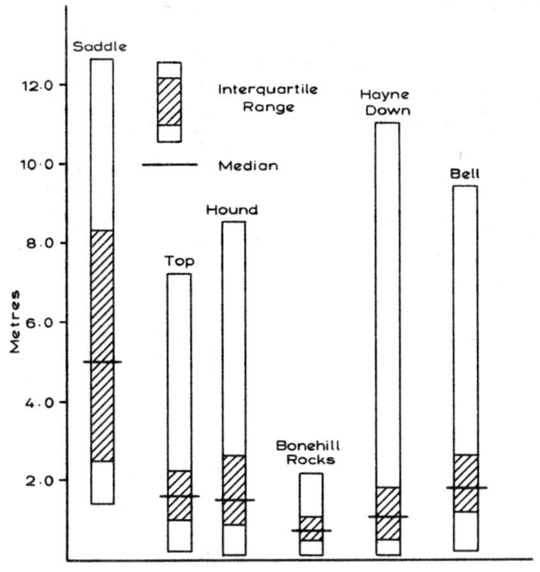

Figure 5.4. Variability of vertical joint spacing for a sample of tors from east Dartmoor.

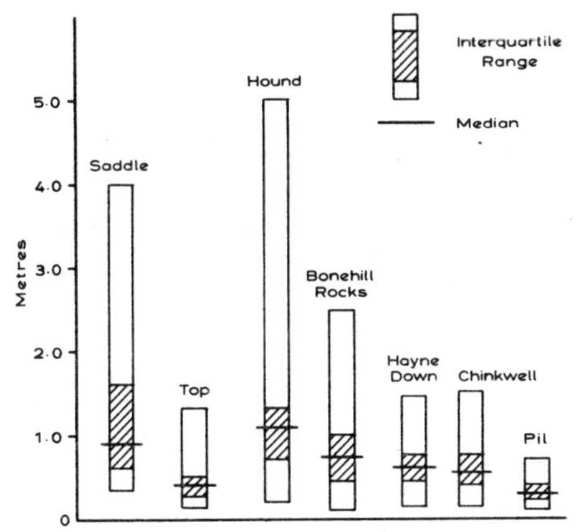

Figure 5.5. Variability of horizontal joint spacing for a sample of tors from east Dartmoor.

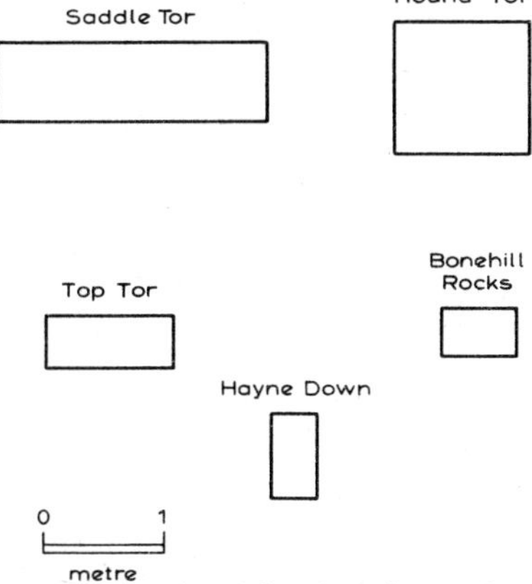

Figure 5.6. Characteristic block sizes based on median spacing of joints.

Figure 5.7. Contrast in joint spacing between fine grained and coarse grained granite.

Figure 5.8. The fine grained granite on Bench Tor has been preferentially weathered to produce a 'rock shelter'.

Nature of joint surfaces

The nature of the joint surfaces dictates the resistance to movement of the blocks. Qualitatively, joint surfaces may be described as smooth, rough or apparently keyed (wavy). Rough and keyed surfaces will give the rock mass greater strength. Waviness has been considered a first order joint wall asperity which would be unlikely to shear off during movement whereas roughness refers to second order asperities which, because they are small, will be sheared off during movement (Anon., 1977). A seven point visual classification, based on the work of Piteau (1970), has been recommended for use when quantitative measurements are not made (figure 5.9).

Category	Degree of roughness
1	Polished
2	Slickensided
3	Smooth
4	Rough
5	Defined ridges
6	Small steps
7	Very rough

Figure 5.9. Suggested categories for the assessment of joint roughness (after Piteau, 1970; Anonymous, 1977).

The nature of thejoint surfaces will also influence the distribution of stresses within the rock mass. The real contact area, because of surface roughness, is only a fraction of the apparent total area and is a function of the applied load. The actual area of contact, expressed as a fraction of the total joint area, is termed the joint contact factor. This figure can be extremely low if blocks are balanced on only a few joint protuberances. Contact stress can be calculated on the assumption that the sum of the loads acting at each zone of contact is equal to the total load on the sample. High contact stresses are common and can attain values considerably in excess of the uniaxial compressive strength of the rock material (Duncan, 1969). Locally, contact zones can be in a state of triaxial stress in which the material on the lower joint surface is prevented from deforming by the material around it. The largest compressive stress possible in such a situation, without plastic yielding, is known as the penetration hardness of the rock, which is usually about three times as great as the yield strength in uniaxial compression.

An interesting relationship exists between the nature of the joint protuberances, the area of contact and the type of stress. The rock composing the protuberances contains voids and microfractures and whilst triaxial stresses are acting, the microfractures remain closed. As the contact area across the joint zone progressively increases, a transformation takes place from a triaxial stress state. Therefore, at some point the confining stresses are inadequate to prevent the opening of vertical microfractures and induced tensile failure occurs.

Ollier (1978) has described rounded corestones of granite split by straight fresh cracks that are not joints of the kind already described. It is suggested that the lower blocks have failed, in the manner described above, because of stresses imposed by the upper boulder or boulders. Ollier has also suggested that the mechanism may be responsible for creating a type of pedestal rock, in which the perched block creates its own pedestal by inducing vertical fractures that reduce the size of the supporting block. Many split boulders have been attributed to the release of internal pressure by unloading (Ollier, 1971) but induced fracture presents another possibility.

It has been estimated that a 10 m diameter granite boulder must rest on an area of only 0.1 m^2 (0.36 m diameter circle) in order to achieve fracture. An even greater force may be required because if the force is applied slowly, the stress may be partially accommodated by plastic deformation. Alternatively, the rock may break at a lower stress if the stress is applied over a long time, because of static fatigue (Scholz, 1968; Martin, 1972). Also, weathering is more effective when the minerals are under stress (stress corrosion) and will help to enlarge cracks. Static fatigue of quartz is also possible due to stress corrosion cracking (Scholz, 1972). Therefore, induced failure should be recognised as a potentially important process in granitic rocks.

Joint or aperture thickness

Joint thickness is defined as the closure which must occur for the joint surfaces to be perfectly in contact and is the equivalent of the maximum cavity depth within the joint zones. The thickness of the joint zones will influence the ease with which water and weathering agents can enter the rock. It will also govern the extent to which deformation of the rock mass may occur under an applied load without the main rock material being affected. Joint thickness is a function of the size of joint surface irregularities and it has been suggested that joint thickness should be measured and described using the terms in figure 5.10.

Joint infill material

Joint infill material may have a decisive effect on the strength of granite. Infill material is either washed down into the rock mass from the surface or created in situ by the weathering or alteration of the rock. Clay material will, in general, produce low strength values unless interlocking of joint faces requires shearing of wall rock during failure. The critical mechanical parameters are the

Description	Joint thickness
Wide	>200 mm
Moderately wide	60 - 200 mm
Moderately narrow	20 - 60 mm
Narrow	6 - 20 mm
Very narrow	2 - 6 mm
Extremely narrow	0 - 2 mm
Tight	Zero

Figure 5.10. Suggested descriptive categories for joint thickness (after Anonymous, 1977).

frictional resistance at the contact between the infill and both the overlying and underlying rock material and the shear strength (cohesion and angle of shearing resistance) of the infill material. Cementation of the joint, such as by quartz or tourmaline, may produce joint properties that are as good as, or even better than, the properties of the rock.

Joint water

Water present in the joints and in the joint infill material will influence the strength of the rock. The removal of infill material by percolating water will reduce the area of contact across the joint zones, increase contact stresses and may result in failure. Joint water may also cause internal erosion producing settlement of the rock mass and incipient joints may open up under the action of water. Where clean fractures exist in the rock mass, water may increase or decrease the angle of frictional resistance. The frictional coefficients of massive crystal structures, such as quartz, increase when in the presence of water, but the anti-lubrication effect diminishes as the roughness of the surface increases.

Joint continuity or persistence

Rock with discontinuous joints will be stronger than rock with continuous joint systems because, for failure to occur, there must be shearing through rock material. Continuous joints create extensive zones of weakness. Information on joint continuity will provide some measure of the percentage of the rock which will be sheared during failure. The degree of continuity or discontinuity can be measured by the distance of joint penetration, although this can only be assessed where the rock is exposed.

An analysis of joints on any small outcrop will illustrate the variety and complexity of joint types, joint openings, infill and continuity that can exist. Figure 5.11 is one such example from an exposure of granite in Brittany.

ANGLE OF DIP	DIP ORIENTATION	JOINT WIDTH	JOINT SURFACE	CONTINUITY	JOINT INFILL	JOINT SPACING
69.5°	315°	TIGHT	R	D	-	0.12 m
71.5°	310°	3 mm	R	C	broken granite	0.04 m
84.0°	30°	2 mm	S	D	-	0.24 m
76.5°	315°	1 mm	S	D	-	0.12 m
74.0°	45°	0.1 mm	K	D	-	0.10 m
76.0°	315°	1.5 mm	S	D	broken granite	0.05 m
72.0°	315°	2.0 mm	S	C	-	0.03 m
80.5°	225°	1.5 mm	K	D	-	0.35 m
85.5°	315°	0.5 mm	S	D	-	0.06 m
83.0°	315°	1.0 mm	R	D	-	0.10 m
80.0°	300°	1.0 mm	S	D	fines	0.23 m
86.5°	300°	1.0 mm	R	D	fines	0.3 m
63.0°	210°	1.0 mm	S	D	-	0.09 m
81.5°	310°	2.2 mm	R	D	fines humus	0.01 m
77.0°	310°	TIGHT	R	D	-	0.02 m
74.0°	20°	TIGHT	S	D	-	0.04 m
82.0°	20°	TIGHT	S	D	-	0.03 m
90.0°	310°	3 mm	S	D	weathered rock	0.06 m
86.5°	130°	4 mm	S	C	weathered rock	

Key
R - rough
S - smooth
K - keyed
C - continuous
D - discontinuous

Figure 5.11. Description of joints on a horizontal granite surface in Brittany.

Figure 5.12. Horizontal joints vary in their degree of openness.

Figure 5.13. Contrast in the openness and continuity of joints.

SPECIFIC GRAVITY	POROSITY %	ELASTIC MODULUS GN/m^2	STRENGTH COMPRESSIVE MN/m^2	STRENGTH TENSILE MN/m^2	REFERENCE
-	-	24.4	62.8	-	Salas, 1968
2.62	0.8	64.1	-	8.62	Nesbit, 1960
2.61	1.7	45.1	-	2.62	Nesbit, 1960
2.63	1.0	73.8	-	-	Nesbit, 1960
-	-	70.7	324.0	-	Hoskings and Horino, 1969
2.74	3.6	69.6	91.7	-	Ruiz, 1966
2.72	3.2	65.8	107.2	-	Ruiz, 1966
2.81	2.2	64.1	111.3	-	Ruiz, 1966
2.65	1.2	75.5	127.1	-	Ruiz, 1966
2.63	1.6	32.5	148.8	-	Balmer, 1953
2.63	1.0	26.2	72.2	-	Balmer, 1953
2.61	2.4	8.96	64.8	-	Balmer, 1953
2.61	2.4	7.79	56.9	-	Balmer, 1953

Figure 5.14. Uniaxial stress-strain parameters for a variety of granites (following Kulhawy, 1975).

SPECIFIC GRAVITY	POROSITY (%)	COHESION $c(MN/m^2)$	FRICTION ANGLE	RANGE OF CONFINING PRESSURES MN/m^2	REFERENCE
-	0.4	55.2	47.7	0.1 - 98	Mogi, 1964
2.66	0.2	55.1	51.0	0 - 68.9	Schwartz, 1964
2.69	0.3	22.1	52.0	0 - 27.6	Stowe, 1969

Figure 5.15. Triaxial stress-strain parameters for a variety of granites (following Kulhawy, 1975).

The same situation occurs on the exposed granites of south-west England. Some joints are wide open whereas other incipient joints are still being formed (figure 5.12). Joint continuity is also highly variable, both vertically and horizontally (figure 5.13). All these factors will influence the strength of the rock.

THE STRENGTH OF GRANITE

The landforms that exist on any rock reflect the interaction of the strength of that rock and the processes fashioning the landform. The mechanical behaviour of rock under stress is, therefore, relevant to an understanding of longterm landscape evolution. One of the most catastrophic rock slides in Norway was caused by the sudden descent, in September 1936, of sheets of granitic rock, with a volume estimated at just over 750 000 cubic metres. The stability of slopes on granite can, therefore, be hard to predict (Terzaghi, 1962, a and b).

Unjointed and unweathered granite is a strong rock with high compressive and tensile strengths, low porosity and high cohesion values (figures 5.14 and 5.15). But granite is normally jointed and it is the joints that determine the strength and behaviour of the rock. Peak shear strength is largely dependent on the effective normal stress acting across the joint but is also sensitive to the degree of surface roughness, the compressive strength of the rock, the degree of weathering, mineralogy and presence or absence of water. The ratio of peak to residual shear strength is dependent on the magnitude of the normal load, the size of joint asperities and the presence or absence of filling material.

The relationships between the physical properties of a jointed rock mass and the mechanics and geometric characteristics of joints can be evaluated in three ways. The first approach is the development of analytical or mathematical models and here the finite element method has yielded a powerful tool for the analysis of the structures in rock (Ghabonssi et al., 1973). The finite element method makes an integrated analysis feasible and the interaction between different rocks and other materials in the structure can be established. The method also allows advanced mathematical models of rock specimens to be the building blocks in a representation of a jointed rock mass. The traditional approach has been to reduce the strength of the rock by a selected amount to take account of the influence of joints. This possesses the disadvantage that the rock mass is still treated as a continuum. A more realistic approach is to treat jointed rock as an aggregate of massive rock blocks separated by joints with special and relevant properties. To aid this methodology, a unit rock block has been defined, which is the smallest homogeneous rock unit produced by a system of joints intersecting a rock mass (John, 1962). Finite element analysis then consists of replacing a complex structure by an assemblage of small elements.

The second approach is the construction of physical models to approximate the behaviour of jointed rock. Simplifications must be made, but such models have the advantage that variables, such as joint inclination and spacing, can be changed over a wide range. Experiments with physical models have shown that when failure occurs by fracture through intact material, rather than along the joints, the strength of the specimen is less than that of an unjointed specimen and the difference increases as the number of joints intersected by the failure surface increases (Elenstern and Hirschfield, 1973). When joints are favourably inclined for potential failure, the strength of the rock is at a minimum and is the strength of the joint surfaces. The third approach is to observe the behaviour of prototypes in the field but this is extremely difficult to do on a large scale.

There are many types of joints and, although detailed quantitative data on the mechanical behaviour of all types have not been obtained, the following characteristics are widely established (Goodman et al., 1968).

a) Joints are tabular. They more closely resemble an irregular line than a zone of appreciable thickness.
b) Joints have essentially no resistance to a net tension force directed in the normal.
c) Joints offer a high resistance to compression in the direction of the normal but may deform under normal pressure, particularly if there are crushable irregularities or compressible filling material.
d) At low normal pressures, shearing stresses along a joint with low inclination create a tendency for one block to ride up on the asperities of the other. Shear strength is largely frictional (Jaeger, 1959, Paulding, 1970). At high normal pressures, shear failure along joints necessitates shearing through the asperities or irregularities. Shear strength is then a function of friction and cohesion created by filling material or interlocking irregularities (Byerlee, 1967).
e) Small shear displacements occur as shear stress builds up below the yield shear stress.

Using these characteristics, the strength of joint surfaces can be subdivided into three joint parameters: the unit stiffness across the joint (kn), the unit stiffness along the joint (ks) and the shear strength (s) along the joint described in terms of cohesion and angle of shearing resistance (Goodman et al., 1968). It may be necessary, under certain circumstances, to introduce an off-diagonal stiffness. The value of the unit stiffness across the joint (kn) depends on:

a) the contact area ratio
b) the perpendicular aperture distribution
c) the properties of the infill material

Unit stiffness along the joint (ks) depends on:

a) the roughness of the joint walls
b) the tangential aperture distribution
c) the properties of the infill material

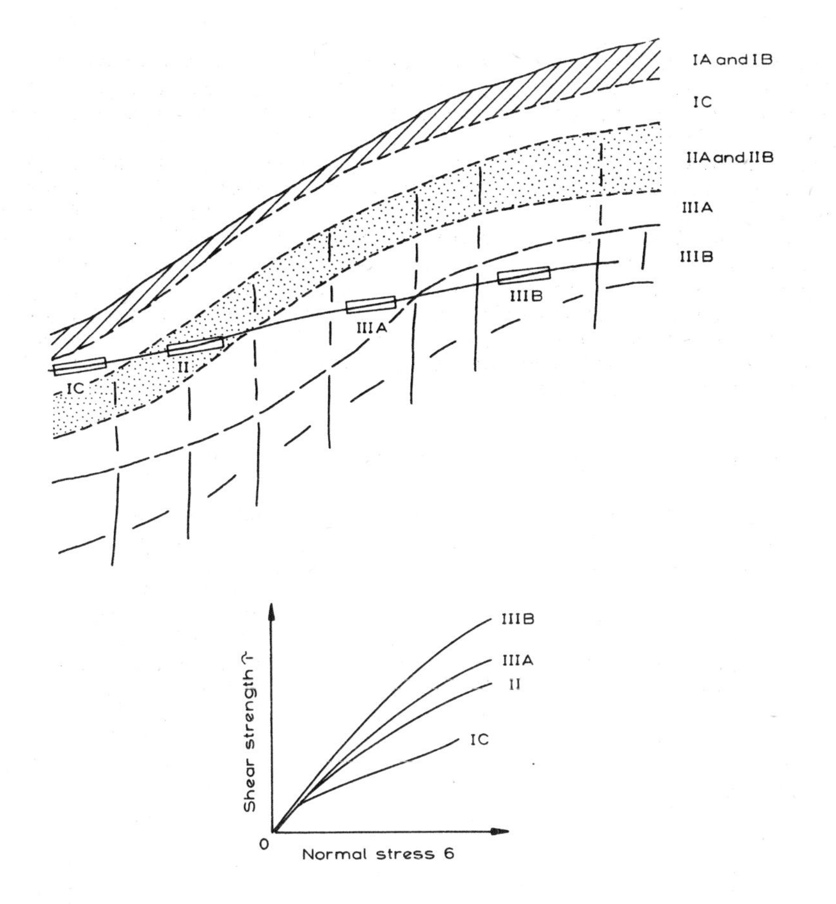

Figure 5.16. Generalised relationship between shear strength and weathering for granite joints (after Deere and Patton, 1971).

The shear strength along the joint depends on:

a) the friction along the joint
b) the cohesion due to interlocking
c) the strength of the infill material

The early studies by Goodman et al. (1968) assumed linear behaviour for kn and ks, except when either tension or shear failure developed. But subsequent work by Goodman (1969) has shown that this assumption is only valid for incipient joints and clean, smooth fractures. Rough fractures and infilled joints exhibit increasing non-linearity. Thus, Henzé et al. (1971) have modified the analysis by using an iterative perturbation solution in which the stiffness values vary with the stress and displacement states.

A hypothetical model for the shear strength of rock joints within a weathering profile has been developed by Deere and Patton (1971) (figure 5.16). The shear strength of fresh rock is controlled by the angle of shearing resistance of the rock material and the roughness of the surface. This produces high angles of apparent shearing resistance at low stress levels due to the tendency of one block to rise up on the asperities. The asperities are sheared off at higher stress levels resulting in an angle of shearing resistance that is close to that of the rock material, although with a large apparent cohesion. As the rock is weakened by weathering, asperities will be sheared off at lower stress levels and the apparent cohesion is reduced for high stress levels. Values for the basic angle of friction are available which allows preliminary strength estimates to be made (figure 5.17). Information is also available for infilled joints and emphasises the drop in shear strength values.

NATURE OF JOINT SURFACE	Ø' (DEGR)	SOURCE
Fresh, fine grained	29 - 35	Coulson 1971
Fresh, coarse grained	31 - 35	Coulson 1971
Discoloured, smooth	29 - 33	Richards 1976
Discoloured, rough	39	Richards 1976
Discoloured, rough	62 - 63	Serafim and Lopez 1961
Weakened, rough	38	Richards 1976
Weakened, rough	45 - 57	Serafim and Lopez 1961
Infilled, stiff silty clay	31	Richards 1976
Infilled, manganese stained	26	Richards 1976
Infilled, kaolin	16.5	Richards 1976
Kaolin-rock contact	12 - 22	Kanji 1970
Illite-rock contact	6.5 - 11.5	Kanji 1970
Montmorillonite-rock contact	4 - 11	Kenney 1967

Figure 5.17. Friction angle data for joints in weathered granite (following Dearman, Baynes and Irfan, 1968).

Relatively simple relationships also exist between fracture spacing and factors such as point load strength and uniaxial compressive strength. These relationships have been combined to produce a chart for assessing ease of excavation which can also be used to infer relative resistance to physical erosion processes. The joint spacings for Dartmoor tors, reported earlier, fall into the higher strength categories. But this simple classification cannot take into consideration the detailed three dimensional geometry of the joint systems, the angle of dip of the joints or the joint infill material.

On the basis of these deliberations it is possible to put into graphical form the factors that are important in governing the behaviour of jointed rock and the shape of the resulting landforms (figure 5.18). It also directs attention to the features that need to be examined, in the field, if an accurate synthesis of landscape evolution is to be obtained. The final part of this essay is an examination of some of these points with respect to the evolution of tors on Western Dartmoor and eastern Bodmin Moor.

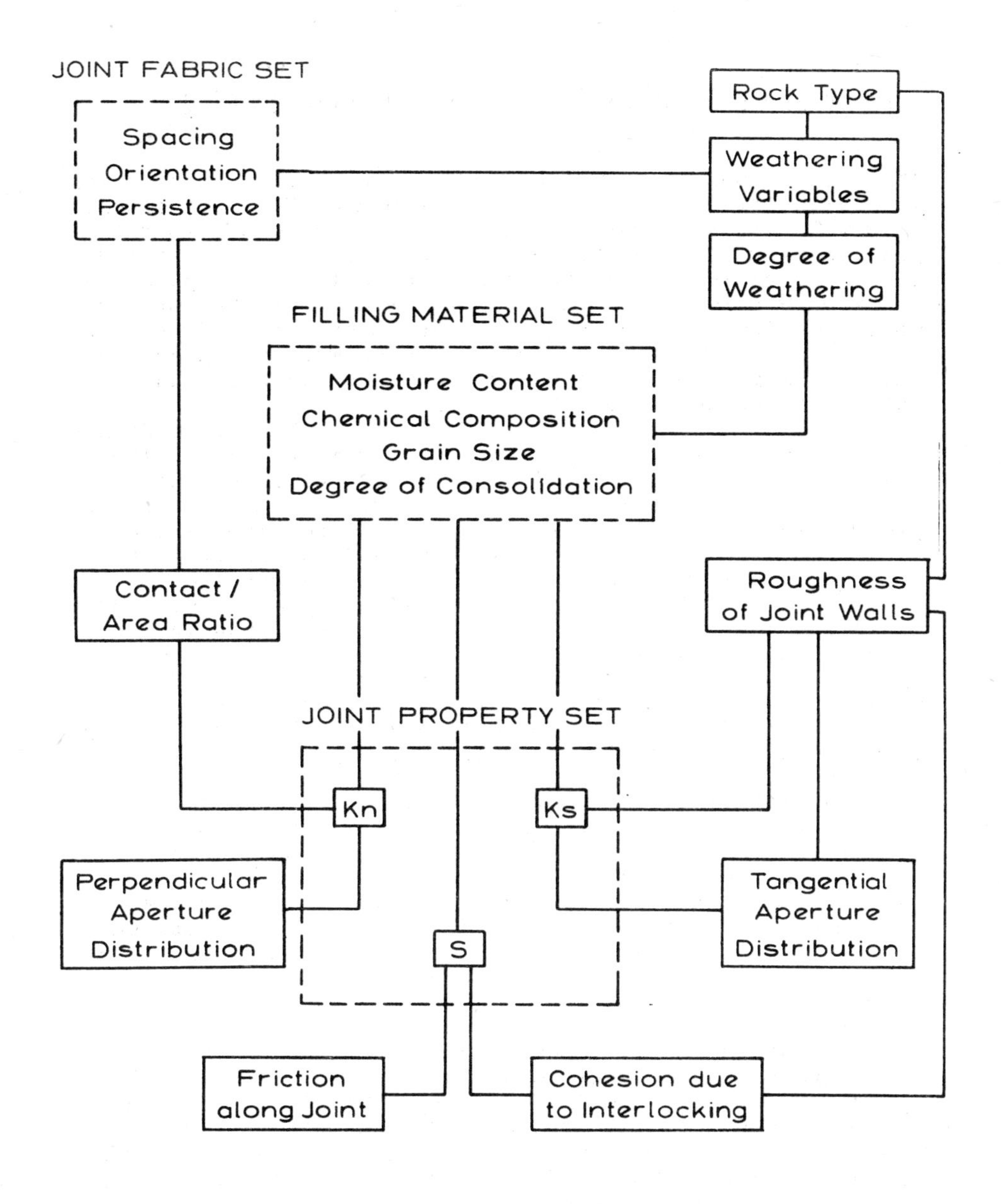

Kn = unit stiffness accross the joint

Ks = unit stiffness along the joint

S = shear strength along the joint

Figure 5.18. Some of the interrelated factors in governing the behaviour of jointed rocks.

TORS OF WESTERN DARTMOOR AND EASTERN BODMIN MOOR

The area of western Dartmoor considered is that surrounding the valley of the river Walkham, from Roos Tor in the north to Pew Tor in the south. This area is very close to the granite margin, with the junction between the granite and the metamorphic aureole passing through the col separating Cox from Staple Tors. The granite exposed on the tors is mostly medium- to coarse-grained and much affected by veins of aplite. The aplite has proved to be more resistant than the medium-grained granite and stands out as knobs and buttresses on the faces of the tors.

The eastern part of Bodmin Moor, stretching from Hawk's Tor to the Cheesewring, presents many similarities. The granite is essentially coarse-grained in which porphyritic orthoclase feldspar is abundant with biotite and muscovite present in almost equal proportions. Tourmaline is common, whilst aplites, quartz-tourmaline veins and quartz veins also occur. There are also areas where the granite is medium-grained, less abundant in feldspar phenocrysts and biotite and with tourmaline more conspicuous than in the coarse-grained variety (Ghosh, 1927).

In both areas there is considerable local accordance between valley and vertical joint directions, especially in the Twelve Mens Moor area of Bodmin Moor. Here a remarkable rectilinear pattern of tors, ridges and associated valleys and depressions occurs in accord with the direction of the vertical jointing. The tors themselves, especially Kilmar, Trewartha and Bearah Tors, are ridge-like in complete contrast to the more conical neighbouring tors of the Cheesewring and Sharp Tor. This may be explained by the different spacings of the vertical joints. The east-west alignment of the tors is favoured by the narrower spaced east-west joints making rockwall retreat from the north and south more possible. The amount of clitter on these slopes attests to the amount of rock retreat from these directions.

It was suggested in an earlier paper that Dartmoor tors can be classified into summit tors, valley-side and spur tors, and small emergent tors found outcropping on the flanks of low convex hills (Gerrard, 1974). In western Dartmoor, Great Staple Tor, Roos Tor, Great Mis Tor and Kings Tor are summit tors. On Bodmin Moor, Sharp, Bearah, Kilmar, Trewartha Tors and the tors of the Cheesewring are of this type. Valley-side and spur tors are often found at the break of slope between the upper convexity and the steeper valley sides e.g. Vixen Tor, Ingra Tor and Hucken Tors on Dartmoor. The amount of stream incision is less on the eastern side of Bodmin Moor but Tregarrick and Hawk's Tors show many of the features of valley side tors. Also, the individual outcrops at the end of Trewartha Tor, such as King Arthur's Bed, fall into this category. The third group are less impressive in terms of dimensions and merely emerge from the flanks of low convex hills e.g. Feather and Little Staple Tors on Dartmoor and Newel and Hill Tors on Siblyback Moor together with the small tors that occur on the northern flank of Craddock Moor. Newel Tor is composed

Figure 5.19. Configuration of joints exposed in the Cheesewring Quarry, Bodmin Moor, Cornwall.

of three small masses, about 3 m high, on a gentle whaleback ridge. Joints are widely spaced.

It is suggested here that all three types, although possessing different features and occurring in contrasting positions, are closely related and may represent a continuum in tor evolution. The connecting thread is provided by the gradual release of the primary confining pressures as incision, weathering and denudation progress. Incision by streams into the domes and ridges would permit the relaxation of stresses still present in the granite and allow further joints to open up. Wahrhaftig (1965), working in California, has noted that orthogonal jointing is present in many of the domes but only as faint cracks which have yet to be opened up. The vertical joints exposed in the Swit Quarry (SX 250724), on Bodmin Moor, are wide apart and very tightly closed whereas those in the neighbouring Cheesewring quarry are more frequent and more open (figure 5.19). The two quarries are in contrasting positions. Swit Quarry has exposed granite low down on the side of Craddock Moor whereas the Cheesewring quarry has almost cut the Cheesewring dome in half.

Jointing opened up in this manner would be most intense in summit areas which may help explain the frequency of jointing on many summit tors, noted earlier. Some confirmation of this is provided by the Cheesewring quarry where the vertical jointing is most intense in the summit region and is almost lacking on the flanks. The major joints that have given the exposed masses their form can be seen to be continuing some way into the granite. These features may help to explain the 'avenue-like' features of many summits.

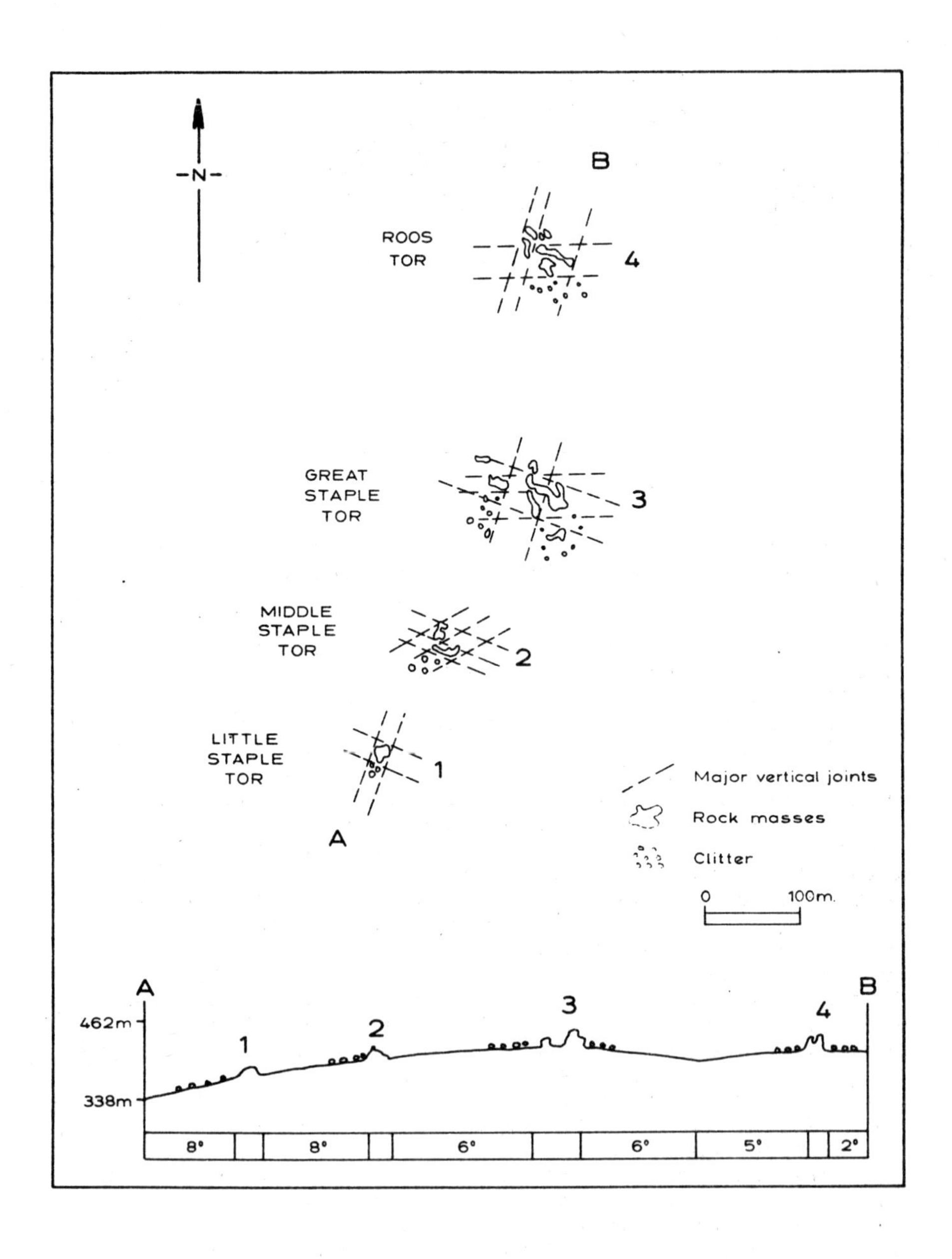

Figure 5.20. Alignment of tors with jointing on the Great Staple Tor ridge.

Many of the larger tors are characterised by missing central portions. In some instances piles of boulders appear to emanate from these avenues giving the impression of preferential tor destruction and removal. Great Staple Tor is the classic example, where the 'avenue' and the series of tors on the one ridge are aligned with the dominant vertical joints (figure 5.20). It has always been assumed, but not proven, that the 'avenues' are the sites of closer spaced joints. The features of the Cheesewring Quarry seem to substantiate this.

Some of the gently dipping joints exposed in Cheesewring Quarry may be of primary origin. The nearness of the granite-country rock contact would make this feasible. But the joints could also be secondary features and the quarries clearly demonstrate how the joints become wider apart with depth and gradually disappear.

These ideas can be supported by an analysis of the intensity of jointing in the three types proposed (Gerrard, 1978). The summit tors are generally massive in type with a network of closely spaced horizontal and vertical joints. In contrast, emergent tors show few vertical joints and those that do occur are widely spaced (figure 5.21).

TOR TYPE	MODAL SPACING HORIZONTAL	JOINTS (m) VERTICAL	MEAN MAXIMUM HEIGHT (m)	MEAN MAXIMUM SLOPE ANGLE (DEGREES)
Summit	0 - 0.99	1 - 1.99	21.8	7.2
Valley side	0 - 0.99	2 - 2.99	26.5	10.4
Emergent	0 - 0.99	>3	9.9	6.3

Figure 5.21. Relationship between jointing, tor type, tor height and topography (adapted from Gerrard (1978)).

The valley side and spur tors are generally intermediate between the two; the intensity of jointing being a function of the amount of incision peculiar to each tor. Thus, Vixen, Ingra and Tregarrick Tors, whose average vertical joint spacing makes them similar to summit tors, possess steep slopes at their base. Ingra Tor, in particular, has been incised on three sides by the river Walkham and two of its tributaries. The difference in size between summit and valley-side or spur tor types is not striking but emergent tors are all of much smaller dimensions.

Representative cross profiles of some of these tors are shown in figure 5.22. There is some tendency for the dominant vertical joints on valley side tors to be parallel to the contours. Thus, on Hucken Tors the only prominent vertical joints occur orthogonal to the steepest slope. It even appears as if the rock masses have moved slightly downslope across the horizontal joints, a phenomenon that has been noted in other granite areas (e.g. Balk, 1939, Terzaghi, 1962b). Widely spaced vertical joints seem to be associated with small tors with gentle slopes. Summit tors

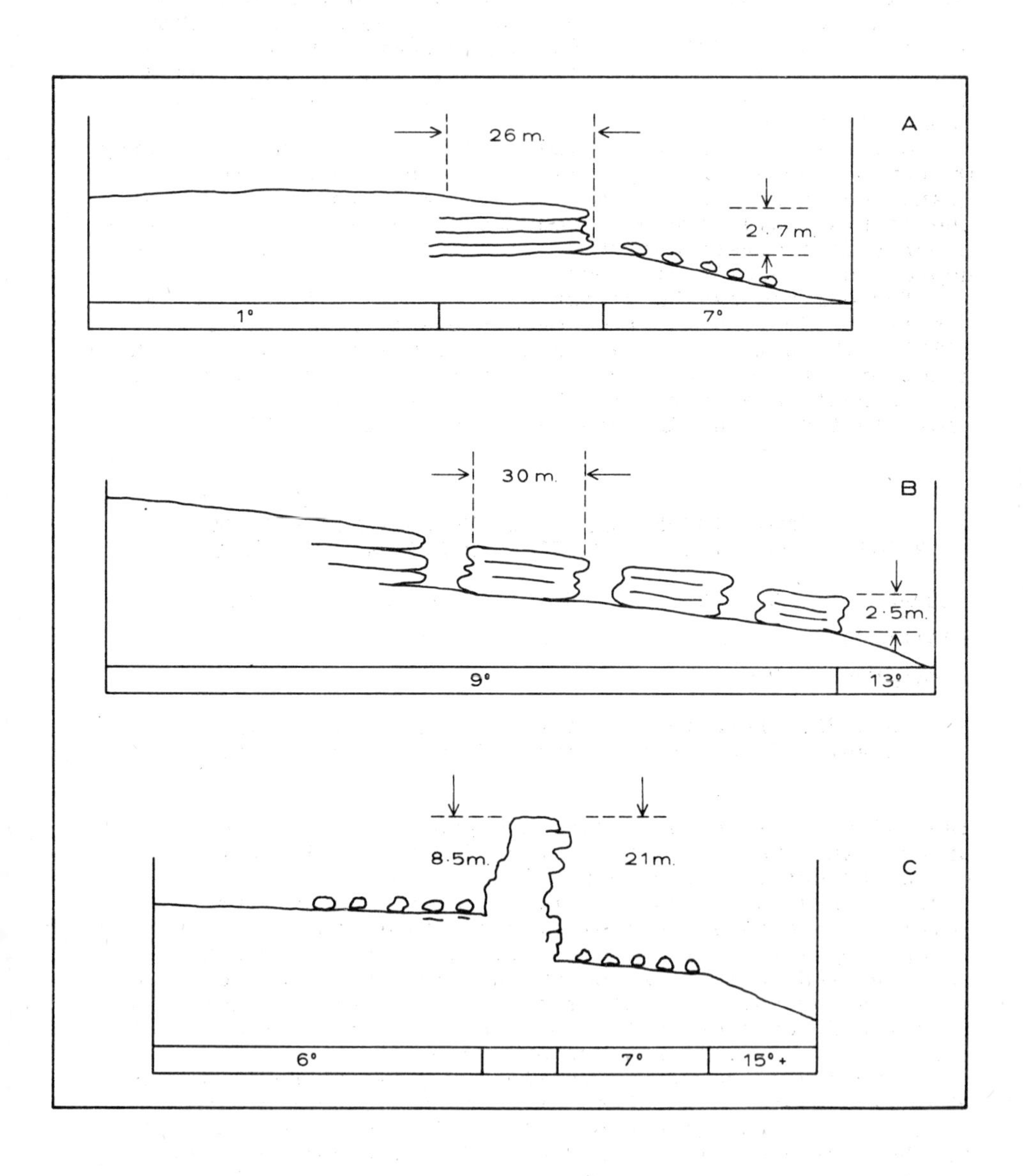

Figure 5.22. Representative profiles of contrasting tor types.

are larger, exhibit much greater complexity of form and possess closer spaced vertical jointing.

It is possible now to consider the way in which the tors and the landscape may have evolved. Initially areas of closer spaced jointing would have been zones of rock weakness and been the sites for increased weathering. This would lead to the compartmentalising of the landscape into positive and negative areas as envisaged by Waters (1957). These zones of closer spaced joints have largely been speculations but recently Knill (1972) has shown that vertical joints with a mean separation of only 0.5 m exist in some valley floors. The ridges and domes would still be essentially in a state of compression, the joints would remain closed and be more resistant to weathering.

As relative relief increases, the joints on the domes would open and weathering would then attack these areas and create the basis for the present tors. The intensity of weathering may have been sufficient to produce completely weathered convex hills as has been suggested for Hingston Down, Cornwall (Dearman et al., 1976). The removal of the weathered granite then exposes the tors. But this simple two stage process of tor formation can only account for a part of the field evidence. It is necessary to consider other possible relationships between the accumulation and removal of the weathered products. At least three situations may occur. Firstly, the ground surface can be relatively stable leading to deepening of the regolith. Secondly, surface instability may occur leading to a gradual removal of regolith and thirdly, a steady state situation is possible where renewal and removal of material continue at similar rates. Each of these states may exist concurrently at different sites and vary in significance at different times, thus a simple explanation based on weathering followed by removal is not adequate to explain the major elements of the granite landscape, a point that has been made elsewhere (Handley, 1952; Hurault, 1967). The same must also be true when considering the effects of periglacial processes.

A highly variable weathering front combined with localised slope instability can lead to the development of random outcrop patterns on slopes and these outcrops, once exposed, will develop less rapidly than the buried rock and will persist in the landscape. Where tor outcrops may represent random outcrops on hillslopes in a state of dynamic equilibrium their use as indicators of past events must depend on the interpretation of the entire landform complexes (Thomas, 1974). The conclusions reached, similar to those of White (1945), are that it is unrealistic to seek a single explanation for apparently similar tors which may have converged towards an equilibrium form from different origins and with the operation of different processes.

The more complex summit and valley-side tors are apparently located where weathering processes, by virtue of highly variable jointing, have been able to act differentially to produce a variety of tor forms. Steeper slopes also allow material to be removed from the base of tors and ensure that the mechanical behaviour and stability of granite, as a jointed rock, have a direct relevance to the evolution of the landscape. The differential joint spacings on Kilmar and Trewartha Tors indicate this. The smaller, emergent tors are likely to be chance exposures which, because of wider spaced joints and gentler slopes, have not developed further. The conclusions suggest that a detailed analysis of the structural features of the granite can provide a synthesis with which to view the variety of granite features that exist on Dartmoor and Bodmin Moor.

Conclusions

At the broad spatial scale, the direction and spacing of joints in granite determine the gross configuration of the landscape. But an interpretation based on structural control must consider the possibility of secondary jointing. The detailed and specific morphology of individual granite landforms, such as inselbergs and tors, are governed by the nature of the joint surfaces, joint infill and consideration of rock strength, as well as by joint spacing and orientation. Principles of rock mechanics are, therefore, relevant to an understanding of granite landforms; a point stressed by Yatsu, in his seminal work in 1966.

REFERENCES

1. Addison, K., 1981, The contribution of discontinuous rock mass failure to glacier erosion, *Annals of Glaciology, 2*, 3-10.
2. Anon, 1977, The description of rock masses for engineering purposes. Report by the Geological Society Engineering Group Working Party, *Quarterly Journal of Engineering Geology, 10*, 355-388.
3. Austin, W.G.C., 1960, *Some aspects of the geology of the Carnmenellis area, Cornwall.* Unpublished M.Sc. Thesis, University of Birmingham.
4. Balk, R., 1939, Disintegration of glacial cliffs, *Journal of Geomorphology, 2*, 303-334.
5. Balmer, G.G., 1953, Physical properties of some typical foundation rocks. *United States Bureau of Reclamation, Denver, Colorado, Concrete Laboratory Report, SP39.*
6. Becker, G.F., 1905, Simultaneous joints, *Proceedings of the Washington Academy of Science, 7*, 267-275.
7. Blyth, F.G.H., 1962, The structure of the north-eastern tract of the Dartmoor granite, *Quarterly Journal of the Geological Society of London, 118*, 435-453.
8. Bradley, W.C., 1963, Large-scale exfoliation in massive sandstones of the Colorado Plateau, *Bulletin of the Geological Society of America, 74*, 519-528.
9. Brammall, A., 1926, The Dartmoor granite, *Proceedings of the Geologists Association, 37*, 251-277.
10. Brunner, F.K. and Scheidegger, A.E., 1973, Exfoliation, *Rock Mechanics, 5*, 43-62.
11. Byerlee, J.D., 1967, Frictional characteristics of granite under high confining pressure, *Journal of Geophysical Research, 72*, 3639-3647.
12. Chapman, C.A., 1958, The control of jointing by topography, *Journal of Geology, 66*, 552-558.
13. Cloos, H., 1936, *Einführung in die Geologie*, Berlin.
14. Coulson, J.H., 1971, Shear strength of flat surfaces in rock. Stability of rock slopes. *Proceedings 13th Symposium on Rock Mechanics, Urbana, Illinois, American Society of Civil Engineering*, 77-105.
15. Crosby, W.O., 1893, The origin of parallel and intersecting joints, *American Geologist, 12*, 368-375.
16. Dale, T.N., 1923, The commercial granites of New England, *Bulletin of the United States Geological Survey,* 738.
17. Dana, J.D., 1876, *Manual of Geology*, 2nd ed.

18. Da Silveira, A.F., Rodrigues, F.P., Grossman, N.F. and Mendes, F., 1966, Qualitative characterisation of the geometric parameters of jointing in rock masses, *Proceedings 1st Congress, International Society of Rock Mechanics*, Lisbon, 225-233.

19. Dearman, W.R., Baynes, F.J. and Irfan, Y., 1976, Practical aspects of periglacial effects on weathered granite, *Proceedings of the Ussher Society, 3*, 373-381.

20. Dearman, W.R., Baynes, F.J. and Irfan, T.Y., 1978, Engineering grading of weathered granite, *Engineering Geology, 12*, 345-374.

21. Deere, D.V. and Patton, E.D., 1971, Slope stability in residual soils, *4th Panamerican Conference on Soil Mechanics and Foundation Engineering, San Juan, Puerto Rico, 1*, 87-170.

22. Dixon, H.W., 1969, Decomposition products of rock substances. Proposed engineering geological classification. *Rock Mechanics Symposium, Stephen Roberts Theatre, University of Sydney*, 39-44.

23. Douglass, P.M. and Voight, B., 1969, Anisotropy of granites: a reflection of microscopic fabric, *Geotechnique, 19*, 376-398.

24. Duncan, N., 1969, *Engineering geology and rock mechanics, II*, (Leonard Hill, London).

25. Durrance, E.M., 1969, Release of strain energy as a mechanism for the mechanical weathering of granular rock material, *Geological Magazine, 106(5)*, 496-497.

26. Edmonds, E.A. et al., 1968, Geology of the country around Okehampton, *Memoir Geological Survey, Great Britain.*

27. Elenstern, H.H. and Hirschfield, R.C., 1973, Model studies on the mechanics of jointed rock, *Journal of the Soil Mechanics and Foundations Division, American Society of Civil Engineers, 99, SM3*, 229-248.

28. Emery, C.L., 1964, Strain energy in rocks, in: *State of Stress in the earth's crust*, ed. Judd, W.R., (Elsevier, New York), 235-260.

29. Exley, C.S., 1958, Magmatic differentiation and alteration in the St. Austell granite, *Quarterly Journal of the Geological Society of London, 114,* 197-230.

30. Exley, C.S., 1965, Some structural features of the Bodmin Moor granite mass, *Proceedings of the Ussher Society, 1*, 157-160.

31. Fookes, P.G., Dearman, W.R. and Franklin, J.A., 1971, Some engineering aspects of rock weathering with field examples from Dartmoor and elsewhere, *Quarterly Journal of Engineering Geology, 4*, 139-185.

32. Gerrard, A.J., 1974, The geomorphological importance of jointing in the Dartmoor granite, in: *Progress in Geomorphology*, ed. Brown, E.H. and Waters, R.S., Institute of British Geographers, Special Publication, 7, 39-51.

33. Gerrard, A.J., 1978, Tors and granite landforms of Dartmoor and eastern Bodmin Moor, *Proceedings of the Ussher Society, 4*, 204-210.

34. Ghabonssi, J., Wilson, E.L. and Isenberg, J., 1973, Finite element analysis for rock joints and interfaces, *Journal of the Soil Mechanics and Foundations Division, American Society of Civil Engineers, 99, SM 10*, 833-848.

35. Ghosh, P.K., 1927, Petrology of the Bodmin Moor granite (eastern part), *Mineralogical Magazine, 21,* 285-309.

36. Gilbert, G.K., 1904, Domes and dome structures of the High Sierra, *Bulletin of the Geological Society of America, 15,* 29-36.

37. Goodman, R.E., 1969, The deformability of joints, in: *Determination of the in-situ modulus of deformation of rocks*, American Society for Testing Materials, STP 477, 174-196.

38. Goodman, R.E., Taylor, R.L. and Brekke, T.L., 1968, A model for the mechanics of jointed rock, *Journal of the Soil Mechanics and Foundations Division, American Society of Civil Engineers, SM 3*, 637-659.

39. Griffith, A.A., 1921, The phenomena of rupture and flow in solids, *Philosophical Transactions, Royal Society of London, 221A*, 163-198.

40. Handley, J.R.F., 1952, The geomorphology of the Nzega area of Tanganyika with special reference to the formation of granite tors, *Comptus Rendus, 19th International Geological Congress, Algiers, 21*, 201-210.

41. Harland, W.B., 1957, Exfoliation joints and ice action, *Journal of Glaciology, 3*, 8-10.

42. Heitfeld, K.H., 1966, Rock loosening near the surface, *Proceedings 1st International Congress of Rock Mechanics, Lisbon, 1*, 15-20.

43. Henzé, F.E., Goodman, R.E. and Bornstein, A., 1971, Numerical analysis of deformability tests in jointed rock - joint perturbation and no tension finite element solutions, *Rock Mechanics, 3(1)*, 13-24.

44. Hodgson, R.A., 1961, Classification of structures on joint surfaces, *American Journal of Science, 259*, 493-502.

45. Horn, H.M. and Deere, D.V., 1962, Frictional characteristics of minerals, *Geotechnique, 12*, 319-335.

46. Hoskins, K. and Horino, F.G., 1969, Influence of spherical head size and specimen diameters on the uniaxial compressive strength of rocks, *Report of the United States Bureau of Mines, Investigation, 7234.*

47. Hurault, J., 1967, L'érosion régressive dans les regions tropicales humides et la genèse des inselbergs granitiques, *Etudes Photo-interpretation, 3.*

48. Irfan, T.Y. and Dearman, W.R., 1978, Engineering petrography of a weathered granite in Cornwall, England, *Quarterly Journal of Engineering Geology, 11*, 233-244.

49. Jaeger, J.C., 1959, The frictional properties of joints in rock, *Geofisca Pura et Applicata, 43*, 148-158.

50. Jahns, R.H., 1943, Sheet structure in granites: its origin and use as a measure of glacial erosion in New England, *Journal of Geology, 51*, 71-98.

51. Jennings, J.N. and Twidale, C.R., 1971, Origin and implications of the A-tent, a minor granite landform, *Australian Geographical Studies, 9*, 41-53.

52. John, K.W., 1962, An approach to rock mechanics, *Journal of the Soil Mechanics and Foundations Division, American Society of Civil Engineers, 88, SM 4*, 1-30.

53. Kanji, M.A., 1970, *Shear strength of soil - rock interfaces,* unpublished M.Sc. thesis, University of Illinois.

54. Kenney, T.C., 1967, The influence of mineral composition on the residual strength of natural soils, *Proceedings Geotechnical Conference Oslo, Norwegian Geotechnical Institute, 1*, 123-129.

55. Kieslinger, A., 1958, Restspannung und Entspannung im Gestein, *Geologie und Bauaresen, 24*, 95-112.

56. Knill, J.L., 1972, Engineering geology in reservoir construction in South West England, *Proceedings of the Ussher Society, 2*, 359-371.

57. Kulhawy, F.H., 1975, Stress deformation properties of rock and rock discontinuities, *Engineering Geology, 9*, 327-350.

58. Leeman, E.R., 1958, Some underground observations relating to the extent of the fracture zone around excavations in some Central Rand mines, *Association of Mine Managers, Transvaal and Orange Free State, Chamber of Mines, South Africa*, 357-384.

59. Lewis, W.V., 1954, Pressure release and glacial erosion, *Journal of Glaciology, 2*, 417-422.

60. Linton, D.L., 1955, The problem of tors, *Geographical Journal, 121*, 470-486.

61. McWilliams, J.R., 1966, The role of microstructure in the physical properties of rock, *American Society for Testing Materials, Special Technical Publication, 402*, 175-189.

62. Martin, R.J., 1972, Time dependent crack growth in quartz and its applications to the creep of rocks, *Journal of Geophysical Research, 77*, 1406-1419.

63. Meunier, A.R., 1961, Contribution a l'étude geomorphologique du nordest du Brésil, *Bulletin de la Société Géologique de France, 7 er series, 3*, 492-500.

64. Mogi, K., 1964, Deformation and fracture of rocks under confining pressure: compression tests on dry rock samples, *Bulletin of the Earthquake Research Institute, Tokyo University, 24(3)*, 491-514.

65. Nesbit, J.H., 1960, Laboratory tests of rock cores from Fremont Canyon Tunnel area - Glendo Unit - Missouri River Basin project, Wyoming, *United States Bureau of Reclamation, Denver, Colorado, Laboratory Report, C-945.*

66. Ollier, C.D., 1971, Causes of spheroidal weathering, *Earth Science Reviews, 7*, 127-141.

67. Ollier, C.D., 1978, Induced fracture and granite landforms, *Zeitschrift für Geomorphologie, NF 22*, 249-257.

68. Onadera, T.F., Yoshinaka, R. and Oda, M., 1974, Weathering and its relation to mechanical properties of granite, *Proceedings 3rd Congress, International Society for Rock Mechanics, Denver, 2A*, 71-78.

69. Oxaal, J., 1916, Norsk Granite, *Norges Geologiske Undersøkelse, 76.*

70. Palmer, J. and Neilson, R.A., 1962, The origin of granite tors on Dartmoor, *Proceedings of the Yorkshire Geological Society, 33*, 315-340.

71. Paulding, B.W., 1970, Coefficient of friction of natural rock surfaces, *Journal of the Soil Mechanics and Foundations Division, American Society of Civil Engineers, 96, SM 2*, 385-393.

72. Piteau, D.R., 1970, Geological factors significant in the stability of slopes cut in rock, *Proceedings Symposium, Planning Open Pit Mines, Johannesburg*, 33-53.

73. Richards, L.R., 1976, *The shear strength of joints in weathered rock*, unpublished Ph.D. thesis, University of London.

74. Roberts, J.C., 1961, Feather-fracture and the mechanics of rock jointing, *American Journal of Science, 259*, 481-492.

75. Ruiz, M.D., 1966, Some technological characteristics of 26 Brazilian rock types, *Proceedings 1st Congress, International Society for Rock Mechanics, Lisbon, 1*, 115-117.

76. Salas, J.A.J., 1968, Mechanical resistances, *Proceedings International Symposium, Determination of Properties of Rock Masses in Foundations and Observation of their Behaviour, Madrid*, 115-129.

77. Scholz, C.H., 1968, Mechanism of creep in brittle rock, *Journal of Geophysical Research, 73*, 3295-3302.

78. Scholz, C.H., 1972, Static fatigue of quartz, *Journal of Geophysical Research, 77*, 2104-2214.

79. Schwartz, A.E., 1964, Failure of rock in the triaxial test, *Proceedings 6th Symposium on Rock Mechanics, Rolla, Missouri*, 109-151.

80. Secor, D.T., 1965, Role of fluid pressure in jointing, *American Journal of Science, 263*, 633-646.

81. Selby, M.J., 1971, Slopes and their development in an ice-free, arid area of Antarctica, *Geografiska Annaler, 53A, 2-3*, 235-245.

82. Selby, M.J., 1977, On the origin of sheeting and laminae in granite rocks: evidence from Antarctica, the Namib Desert and the Central Sahara, *Madoqua, 10(3)*, 171-179.

83. Serafim, J.L., and Lopez, J.P., 1961, In situ shear tests and uniaxial tests of foundation rock of concrete dams, *Proceedings 5th International Conference, Soil Mechanics and Foundation Engineering, Paris, 1*, 533-539.

84. Simmons, G., Todd, T. and Baldridge, W.S., 1975, Toward a quantitative relationship between elastic properties and cracks in low porosity rocks, *American Journal of Science, 275*, 318-345.

85. Snow, D.T., 1968, Rock fracture spacings, openings and porosities, *Journal of Soil Mechanics and Foundations Division, American Society of Civil Engineers, 94, SM1*, 73-91.

86. Soen, O.I., 1965, Sheeting and exfoliation in granites of Sermersoq, South Greenland, *Meddelelser om Grønland, 179(6)*, 1-40.

87. Stowe, R.L., 1969, Strength and deformation properties of granite, basalt, limestone and tuff at various loading rates, *United States Army Corps of Engineers, Waterways Experimental Station, Vicksburg, Mississippi, Miscellaneous Paper, C91-1*.

88. Suess, E., 1913, Uber Zerlegung der Gebirgsfoldenden Kraft, *Geologie Gessellschaft Wien, Mittelinger, 6*, 13-60.

89. Tarr, R.S., 1891, The phenomena of rifting in granite, *American Journal of Science, 41*, 267-272.

90. Terzaghi, K., 1962a, Dam foundations on sheeted granite, *Geotechnique, 12*, 199-208.

91. Terzaghi, K., 1962b, The stability of slopes on hard unweathered rock, *Geotechnique, 12*, 251-270.

92. Thomas, M.F., 1974, Granite landforms: a review of some recurrent problems of interpretation, in: *Progress in Geomorphology*, eds. Brown, E.H. and Waters, R.S., Institute of British Geographers, Special Publication, 7, 13-37.

93. Thorp, M.B., 1967, Closed basins in Younger Granite massifs, northern Nigeria, *Zeitschrift für Geomorphologie, NF 11*, 459-480.

94. Twidale, C.R., 1972, The neglected third dimension, *Zeitschrift für Geomorphologie, NF 6(3)*, 283-300.

95. Twidale, C.R., 1973, On the origin of sheet jointing, *Rock Mechanics*, *5*, 163-187.

96. Wahrhaftig, C., 1965, Stepped topography of the southern Sierra Nevada, California, *Bulletin of the Geological Society of America*, *76*, 1165-1190.

97. Waters, R.S., 1954, Pseudo-bedding in the Dartmoor granite, *Transactions of the Royal Geological Society of Cornwall*, *48*, 456-462.

98. Waters, R.S., 1957, Differential weathering in old-lands, *Geographical Journal*, *123*, 503-513.

99. White, W.A., 1945, The origin of granite domes in the south-east piedomont, *Journal of Geology*, *53*, 276-282.

100. Whittle, C.L., 1900, Rifting and grain in granite, *Engineering and Mining Journal*, *70*, 161.

101. Willard, R.J. and McWilliams, J.R., 1969, Micro-structural techniques in the study of physical properties of rock, *International Journal of Rock Mechanics and Mining Science*, *6*, 1-12.

102. Woodworth, J.B., 1896, On the fracture system of joints with remarks on certain great fractures, *Proceedings of the Boston Society for Natural History*, *27*, 163-184.

103. Yatsu, E., 1966, *Rock control in geomorphology*, (Sozosha, Tokyo).

104. Zirkel, F., 1894, *Lehrbuch der Petrogrophie*, 2nd. ed.

6. Man and the Somerset Levels

F.A. Hibbert

1. INTRODUCTION

The area known as the Somerset Levels is a tract of peatland in northern Somerset which lies between the high ground of the Mendip Hills to the north and the Quantock Hills to the south-west (Figure 6.1). Within this basin there are ridges of higher ground, the Wedmore Ridge and the Polden Hills, together with other more isolated lias outcrops and 'islands' of Quaternary sands and outwash material. These latter, which are known locally as the Burtle Beds, have played what appears to have been an important role in the life of prehistoric man in the area. Research into the vegetational history of the area and its associated archaeology has been going on for some forty years and has, in the main, been concentrated in the Vale of Avalon, north of the Chilton Polden Hills and south of the Wedmore Ridge (Figure 6.2).

The area has been under the influence of flooding both by the sea and by fresh water over the last 10 000 years. Early work (Godwin, 1941) showed that a thick deposit of marine clay lies beneath the superficial peat deposits. It was subsequently shown that beneath the clay there lies a further peat bed (Coles, Hibbert and Orme, 1973). Dating of this early peat is rather uncertain but the contact between this and the overlying clay has been dated to 6520 BP.

It is generally found that around 5500 BP the sea receded from the Levels and the build up of biogenic deposits commenced.

2. PEAT STRATIGRAPHY

The presence of peat is due to the persistent waterlogging of the low-lying flat plateau. Under such conditions the

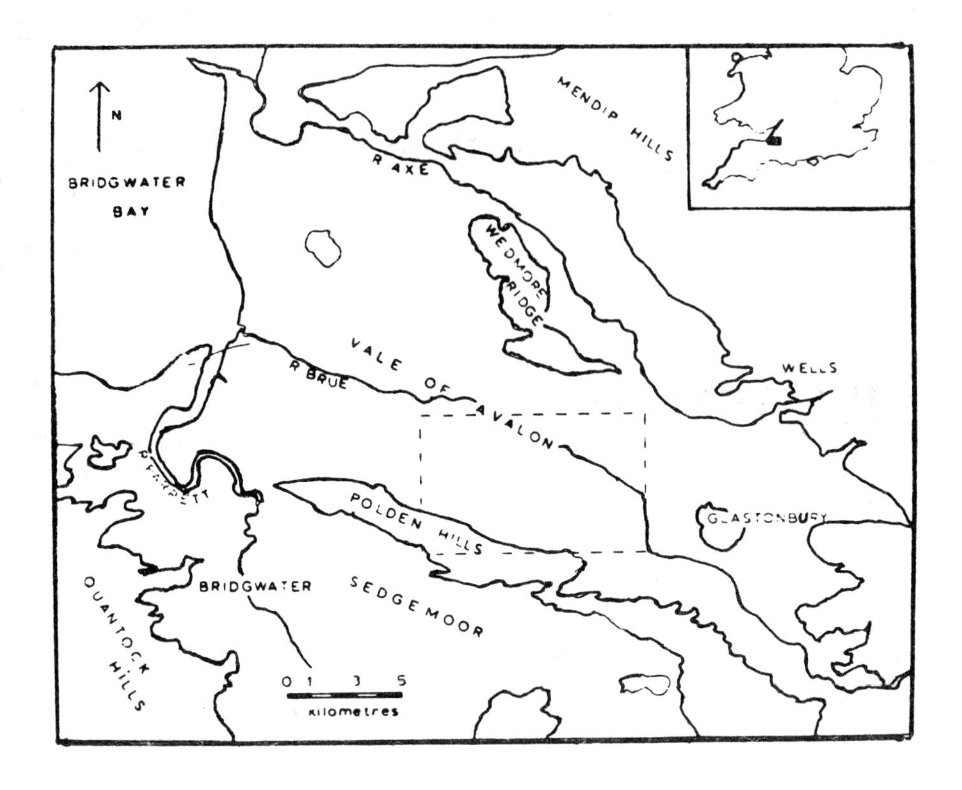

Figure 6.1. Map of the Somerset Levels.

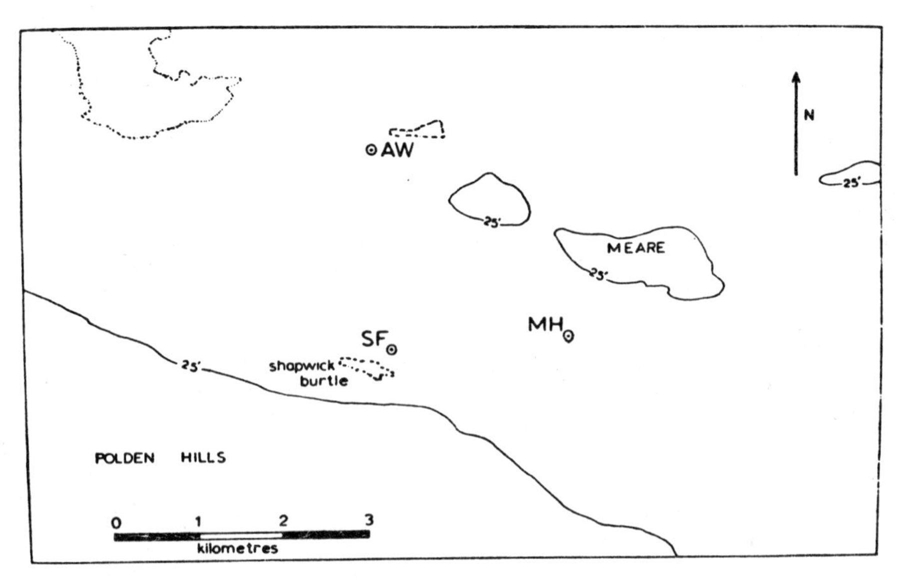

Figure 6.2. The Vale of Avalon showing the location of the three pollen diagrams (AW is the Abbot's Way, SF the Sweet Track and MH the Meare Heath trackway).

normal agencies that lead to the decay of dead organic material do not operate. As a consequence there is a build-up of organic matter in a chronological sequence and, provided that the waterlogging persists, then that sequence is unbroken and provides a continuous record of events which may be investigated.

As the sea level fell the presence of beach deposits to the west of the Levels together with the low-lying nature of the land led to the development of a shallow lake. No doubt this would have been brackish initially, but the plant remains found in the peat are, in the main, indicative of a fresh-water lagoon. The presence of *Phragmites* (the Reed) throughout the deposit tells that the water was never more than about 1 metre deep. However it is likely that there were deeper interconnecting water channels between the isolated areas of shallower water, a situation not unlike the vegetation and topography of the Danube delta area today.

As time passed there was a progressive build up of organic debris that further shallowed the water and so allowed plants more suited to this to become dominant. In particular *Cladium mariscus*, the saw-toothed sedge, a plant which is intolerant of water deeper than 0.5 m and especially intolerant of wildly fluctuating water levels, became present across the whole area. Accompanying *Cladium* are the first regular remains of trees. Birch and Alder were undoubtedly growing in the rafted vegetation as well as in areas that were already exceptionally shallow and therefore were more stable. The plants that were present at this time include many that preferred nutrient-rich water, this indicating that the whole area was under the influence of the ground water table draining from the surrounding high land.

The peat then became dominated by remains of trees. Wood of Birch, Alder or Willow comprises almost the whole of the deposits leaving a layer of wood peat throughout the whole area. This woodland is still fed and maintained by ground water and became established across the levels around 5000-4500 BP.

Following the wood peat there are deposits of quite a different kind. The peat that follows is made up of plants that are quite intolerant of nutrient rich water, rather they tolerate the relatively poor nutrient status of rain water. They are plants such as *Calluna* (Ling) *Eriophorum* (Cotton grass) and *Sphagnum* moss. They show that quite a different ecosystem developed in the area - one that was growing above the water table. For this to happen, and to develop from a fen wood, the rainfall has to be at least 1 metre per annum. Extensive peat deposits of this kind, known as raised-bog peats, are found in the area under investigation and the sort of picture that such deposits present is comparable to the vast raised bogs of mid-Wales or those on the central plain of Ireland. The surface topography is a mosaic of wetter hollows, with quite deep pools, having drier hummocks between.

The peat deposits therefore tell of a three-fold change in the environment of the Levels, initially a shallow, open water lake, followed as the water shallows by the development

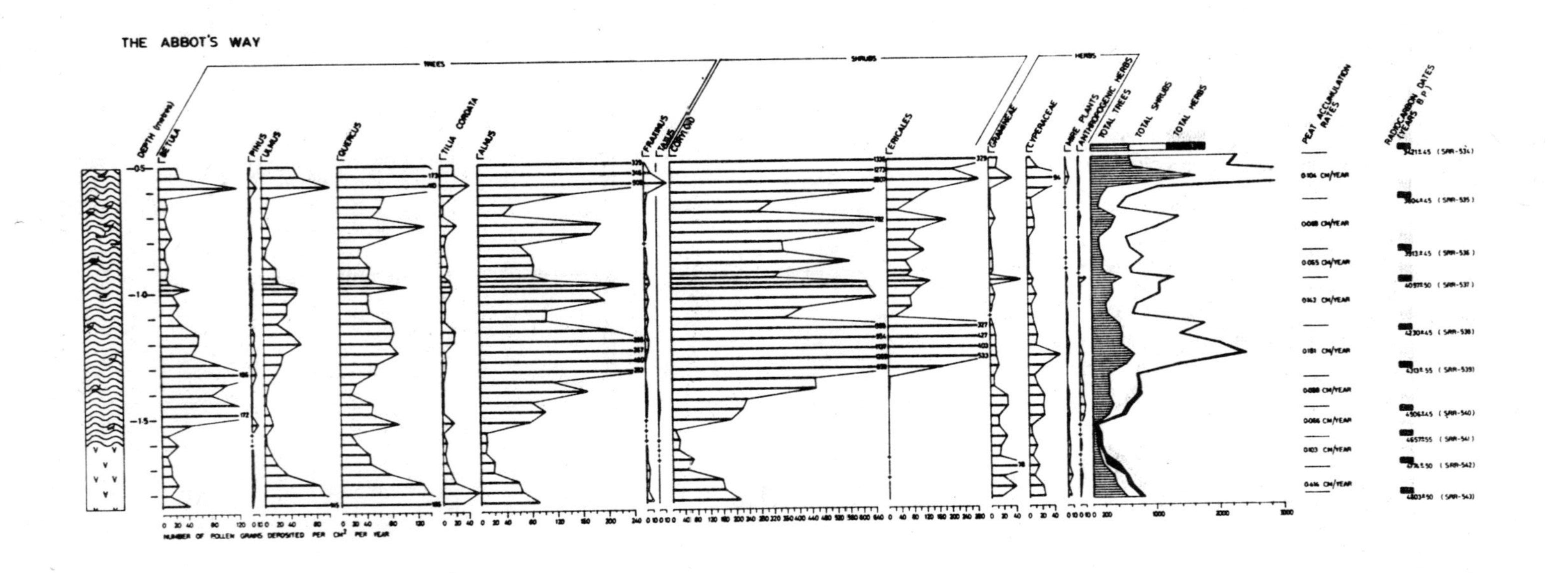

Figure 6.3. Absolute pollen frequency diagram from the Abbot's Way.

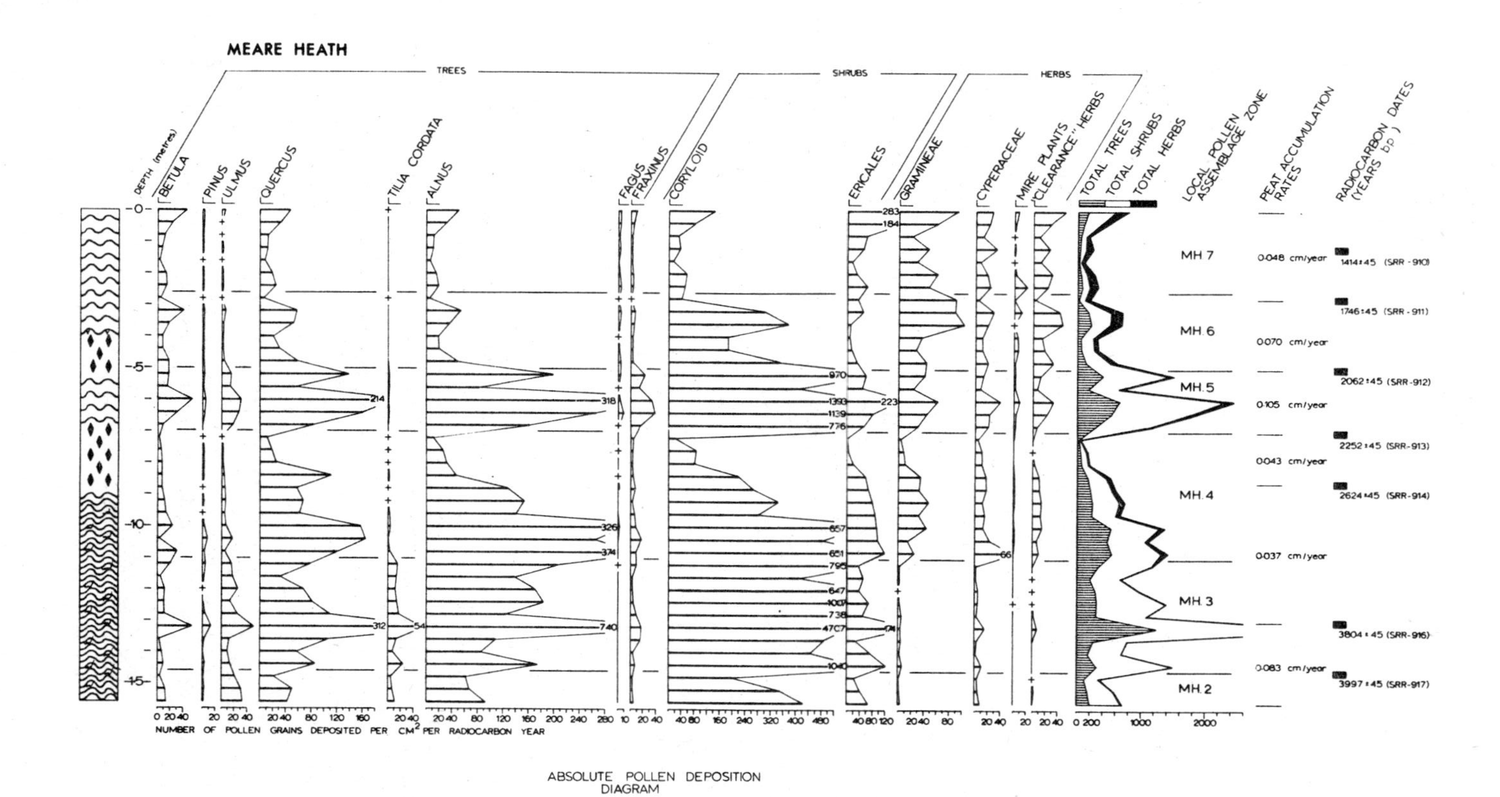

Figure 6.4. Absolute pollen frequency diagram from the Meare Heath trackway.

of fen woodland which, in turn, was followed by the growth of raised bog. This was later interrupted by an oscillation between renewed flooding by ground water and further growth of raised bog from around 2600 BP to about 1800 BP.

3. POLLEN EVIDENCE

The pollen record substantiates the results gained from the analysis of the macroscopic remains of the peat.

Pollen from plants that live in open water is recorded from the lowermost deposits. Sedge pollen increases as the water becomes more shallow and the reed swamp establishes itself. Tree pollen is present throughout the deposits, no doubt coming from the vegetation surrounding the open water and the high land enclosing the basin. As the establishment of fen wood proceeds then there is an increase in the pollen of trees that are present in such woodlands, notably pollen of Birch and Alder. The vegetation of the dry, high ground is shown to be of a mixed-oak forest type with little evidence of the forest cover being broken, pollen of herbaceous plants being represented only by low percentages.

As the peat type changes to that of a raised bog then the pollen record shows an increased representation of such plants as *Calluna* (Ling) and *Empetrum* (Crowberry) together with an increase in the spores of *Sphagnum* (Bog moss). Dry land pollen shows a continued presence of mixed-oak forest but with an increase in the representation of pollen likely to have originated in clearances within the forest cover.

The pollen record is represented by the pollen diagrams from Abbots Way (Figure 6.3) and Meare Heath (Figure 6.4). Earlier formed deposits, resting in the clay, are found at Abbots Way whilst Meare Heath overlaps the later deposits from Abbots Way and carries the record on to the present surface of the peat deposits.

Detailed examination of both these diagrams will show both the direction and scale of the change in the pollen record. The diagrams are drawn upon the basis of pollen influx and are presented as absolute pollen deposition. This removes from the diagram the influences that percentage representation brings in the way that if one taxon increases then, somewhere in the pollen sum used to calculate the percentage frequency, other pollen representation must 'fall' to accommodate this. Such a fall may not be real and an absolute pollen diagram records pollen representation in such a way that this 'competition' for representation is removed. However, the calculation for scarcer taxa is clearly less reliable statistically.

The purpose of this paper is to explore the activity of man within the Somerset Levels and discussion of the pollen diagrams will be restricted to that end. A more comprehensive review of the vegetational history has been reported elsewhere (Beckett and Hibbert, 1979).

4. THE ACTIVITY OF MAN

Many earlier workers have demonstrated the fact that man has been interested in the Somerset Levels from early Neolithic times (Godwin, 1960, Coles and Hibbert, 1968).

The extent of the discoveries depended upon peat cutting in the area and as this has become more extensive, and the speed of extraction much faster, then the rate of discovery also increased. It was earlier thought that the building of trackways across the Levels was determined by the sort of environment that man was facing. As the fen-woodland became more stable, though subject to seasonal flooding - there was a burst of trackway building. Many of these earlier trackways were of a simple kind, bundles of poles held together by pegs laid upon the bog surface (Godwin, 1960). The exception to this being the Sweet Track (Coles, Hibbert and Orme, 1973) which is of a much more complex construction involving massive timbers, of dry land origin and advanced carpentry techniques. In any event they are preserved in peat that by its nature indicates a stable fen woodland, with the possibility of seasonal flooding, a suitable environment for the construction of trackway, that might be expected to survive for a few years depending largely upon the complexity of construction. In addition there are trackways - or platforms - of a third kind where the method of construction is to pre-fabricate hurdles, to carry them to the Levels and use them to facilitate the passage of man (Coles and Orme, 1977).

As the fen wood was replaced by raised bog deposits then the rather impassable conditions of the former are replaced by quite a different environment. Modern raised bogs are passable by man at almost any time of the year providing that he is intelligent enough to avoid the wetter hollows and to keep his pathway to the drier hummocks. It is therefore not surprising that the occurrence of trackways in the peat falls away as raised bog becomes established across the area. One notable exception is the Abbot's Way track, a rather stout trackway comprised, almost entirely, of Alder laid in a transverse configuration. This trackway lies in raised bog peat and this would indicate that a purpose other than the provision of a walkway for man alone was in mind. Exactly what this was remains unsolved and an alternative view might be that it had some significance in ritual. In any event detailed boring across several hundred metres revealed that it did not run in a straight line, rather that it twisted and bent, no doubt to avoid the wetter pools and follow the line of the drier hummocks.

Trackway building had a renewed phase as the conditions on the raised bog became wetter. The Meare Heath trackway, which is dated around 3200 BP is of a complicated construction with massive planks of oak, associated pegs and rails all of which indicate a sophisticated manner of construction. The trackway lies in raised bog peat whilst the peat above it contains the remains of *Cladium* which indicates a return to flooding of the levels by surface water. Whilst the trackway is not contemporary with this flooding horizon it may be assumed that the deterioration in conditions which is reflected by the flooding episode, began somewhat earlier and that the trackway was a response to the flooding of a traditional route across the raised bog.

Godwin (1960) reported other Bronze-age trackways that were associated with this later flooding episode. Many of these have been lost due to peat cutting but enough evidence was recorded to suggest that the flooding of the levels, which is recorded in the peat stratigraphy, is accompanied by a phase of trackway building.

Recent pollen analysis (Beckett and Hibbert, 1979) together with radio-carbon dating has shown that, from early Neolithic times, man has been involved in clearing forest from the high ground around and within the Somerset Levels. The pollen assemblage zones represent periods of forest clearance separated by periods of forest regeneration. The three profiles that contribute towards this study come from Abbots Way, Meare Heath and the Sweet trackway (Figure 6.5).

At around 4700 BP there is a marked decline in the pollen of *Ulmus* (Elm) and an increase in the representation of Gramineae (grasses), *Plantago lanceolata* (Ribwort plantain) and *Artemisia* (mugwort). This is taken to represent an opening of the forest cover and the beginnings of agricultural activity in the clearances created. The increase of grass, plantain and bracken are indicators of a pastoral economy and whilst there are a few scattered occurrences of cereal pollen, the evidence nevertheless points towards a pastoral economy.

The clearance phase seems to have lasted for about 400 years following which time the pollen of *Ulmus* rises again and the pollen of 'weeds', or plants of open ground, falls away. This clearance phase coincides with the construction of many trackways across the levels. The phase of forest regeneration tells of *Fraxinus* (Ash) growing in the abandoned clearings and the re-establishment of mixed oak forest in the area. This phase lasts until 4000 BP when there is evidence of renewed clearance. *Ulmus* pollen falls, yet *Tilia* (Lime) pollen remains almost unchanged and there is little increase in the representation of herbaceous pollen. This would seem to indicate a limited clearance and yet the effects persisted for some 400-600 years. It is possible that the pollen rain reflects a large number of small clearance phases over the region throughout the period. The Abbot's trackway was built during this clearance phase.

There then follows another period of forest regeneration but this was short lived for at around 3300 BP there is a very marked fall in the pollen of *Ulmus* (Elm) and *Tilia* (Lime) together with an increase in pollen of open and disturbed ground. The Meare Heath trackway is associated with this clearance phase and the extent of clearance suggests that Bronze Age man was making greater inroads into the forest of the area than had been undertaken before. Many other trackways were built at this time and the clearance persisted for many hundreds of years up until 3200 BP. At this time there was an expansion of forest cover once again but only briefly, for at 2100 BP there is the beginning of a major clearance of forest, possibly on a

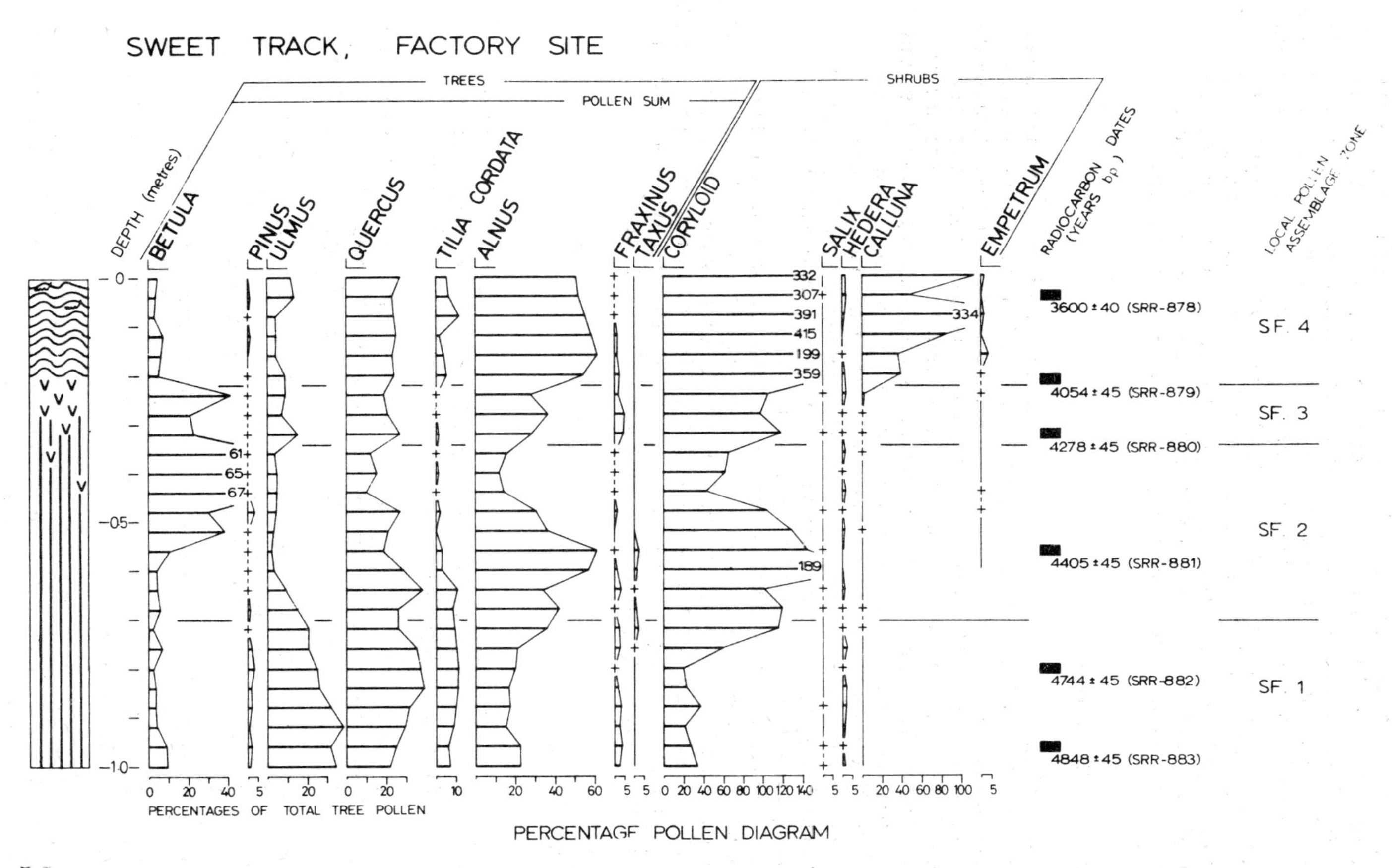

Figure 6.5. Frequency pollen diagram from the Sweet Track.

much wider regional scale, that persisted throughout Romano-British times, certainly beyond 1414 BP, which represents the date of the uppermost peat deposits remaining in the area. No doubt this last clearance phase was due to improved technology and the eventual clearance of Oak from the heavy clay soils of the valleys - for the pollen of all tree taxa fell during and beyond Roman times. In Bronze age and later times there is ample representation of pollen from arable cultivation and pollen of a pastoral economy. No doubt both these activities persisted in the surrounding high ground as the population grew and the need for agriculture to support such a growth was established.

5. CONCLUSION

The unique juxtaposition of trackways and peat in the area of the Somerset Levels allows a correlation to be established between the environment within which these trackways were built and the number that were associated with the needs of man from Neolithic times onwards. The influx of pollen rain from the surrounding high land and the 'islands' within the peat enable conclusions to be drawn as to the nature and extent of the activity of man in the area. Successive clearances, of an increasing intensity, are recorded and periods of trackway building are shown to be associated with the clearance phases.

The picuture that emerges is one in which man has played an important role in modifying the vegetation of the area. Primaeval forest had been cleared over a period of 3000 years and replaced eventually by a system of both arable and pastoral farming on such a scale as to have reduced the forest cover considerably by Anglo-Saxon times.

REFERENCES

1. Beckett, S.C. and Hibbert, F.A., 1979, Vegetational change and the influence of prehistoric man in the Somerset Levels, *New Phytologist*, *83*, 577-600.

2. Coles, J.M. and Hibbert, F.A., 1965, Prehistoric roads and tracks in Somerset, England: 1. Neolithic, *Proceedings of the Prehistoric Society*, *31*, 238-258.

3. Coles, J.M. and Orme, B.J., 1977, Neolithic hurdles from Walton Heath, in: *Somerset Levels Papers*, ed. Coles, J.M. (Stephen Austin, Hertford).

4. Coles, J.M., Hibbert, F.A. and Orme, B.J., 1973, Prehistoric roads and tracks in Somerset, England: 2. Neolithic, *Proceedings of the Prehistoric Society*, *39*, 256-293.

5. Godwin, H., 1960, Prehistoric wooden trackways of the Somerset Levels: their construction, age and relation to climatic change, *Proceedings of the Prehistoric Society*, *26*, 1-36.

7. Channel changes in regulated rivers

G.E. Petts

1. INTRODUCTION

Dams have been constructed for a variety of functions such as flood control, water storage, hydro-electric power generation and river-regulation to provide for downstream demands - irrigation, navigation, domestic and industrial water abstraction, and amenity. Regardless of the objectives for reservoir creation marked changes in the physical characteristics of the downstream areas will result. These changes are diverse in character; the rates of change are highly variable; and alternative pathways can result in a single, particular effect. Furthermore, a single important consequence may arise immediately subsequent to dam closure but others may become detectable only after several years. Thus, it is difficult to formulate a general model to describe the morphological consequences of upstream impoundment.

Attempts to explain the different channel changes induced by apparently similar but spatially independent impacts are rare and have failed to produce a generally applicable model primarily because channel changes have been associated with only a single control variable. The effect of each individual impoundment upon the river channel downstream is produced by a unique combination of climate, geology, project size and operational procedures, etc. so that a wide range of river channel responses are potentially generated by river regulation. Indeed, descriptions of the channel changes induced by headwater impoundment demonstrate that a range of effects may occur consequent upon the application of a particular impact. Wolman (1967), for example, related channel changes to flow alteration for eleven reservoirs in USA, but at a pre- to post-dam discharge ratio of 0.75, channel capacity demonstrated increases of 40 per cent and decreases of 50 per cent. A second problem in making a general assessment of the morphological consequences of river regulation, particularly on large rivers, arises from the variable downstream character of the drainage basin and channel network. Different channel responses have, for example, been observed below Glen Canyon Dam on the Colorado River. Channel bed degradation and scour has occurred for 24 km below the dam (Pemberton, 1975) but further downstream, channel aggradation has dominated (Turner and Karpiscak, 1980). This paper seeks to create a general framework for the examination of diverse channel responses through the application of the geomorphological concepts of 'work' and 'event effectiveness' to an evaluation of the significance of mainstream and tributary discharges within regulated rivers.

2. THE APPROACH

Studies of human impacts upon river systems have generally embodied the concept of magnitude and frequency. Such studies have considered that flood events of moderate magnitude and frequency perform most of the work, defined as the quantity of material transported over a given distance in a given period of time (Wolman and Miller, 1960). The concepts of geomorphic thresholds, complex response and episodic change (Schumm, 1977) have, however, provided an improved approach for the study of river channel adjustments by directing attention towards the effectiveness of individual, high magnitude floods (Wolman and Gerson, 1978; Newson, 1980a) and, particularly, towards the significance of vegetation for channel change. Wolman and Gerson described the effectiveness of an event in forming landscape features in terms of the rate of recovery of channel morphology following alteration by an extreme event. Thus, event effectiveness depends not only upon the absolute magnitude, or force, of a flood discharge and the frequency with which it recurs, but also upon the magnitude and frequency of the constructive or restorative processes during the intervening intervals between 'effective' flood events.

Vegetation, specifically the composition and density of the plant cover, plays an important role in determining both the magnitude and frequency characteristics of effective events and the rate of landscape recovery. The existence of a well established vegetation cover will markedly influence the flood magnitude required to induce any significant change of channel morphology. Thus, Graf (1979) demonstrated the influence of the biomass of vegetation on the valley floor as a threshold in the development of montane Arroyos and gullies in the Front Range of Colorado. The rate of vegetation establishment in a given region is primarily dependent on moisture. If the rate is rapid, then regardless of the river channel changes caused by an individual flood event, and assuming a supply of sediments, the reconstruction of the channel dimensions prevailing prior to the event should also be rapid. Thus, within temperate regions river channels widened by rare flood events have often regained their original form in a matter of months or years whilst under semi-arid conditions rare events may produce progressive change. An effectiveness approach, however, has been primarily directed to the examination of destructive flood events under natural conditions.

Channel response to human impacts often involves the adjustment of the channel morphology from one quasi-equilibrium state to another, imposed, condition. Time lags commonly exist between the application of a stimulus and the initiation of channel change (the reaction time) and the period between the beginning of change and the establishment of a new steady state (the relaxation time). For uninterrupted relaxation paths, theoretical (Allen, 1974, 1977), empirical (Graf, 1977) and experimental (Begin *et al*, 1981) evidence has been provided to suggest that the form of adjustment is described by an initial rapid rate of change followed by a progressively decreasing rate as the new steady state is approached. In reality, however, the relaxation path represents a complex response characterised by alternating phases of erosion and deposition. The form of the relaxation path will reflect the magnitude and frequency characteristics of destructional and constructional processes, in relation to the rate of vegetation establishment.

Reservoirs markedly alter the processes operating in the downstream river system: the retention of water behind a dam and its gradual release downstream result in the reduction of peak discharges, the regulation of the flow regime, and the isolation of the sediment sources. However, the significance of these process changes for a discrete channel reach is dependent upon the character of the processes - reflecting the interaction of both mainstream and tributary sources - and of the channel morphology prior to headwater impoundment (figure 7.1). River regulation commonly represents a compromise situation between two conflicting extremes. Flood regulation requires the provision of 'empty space' so that the reservoir storage should be maintained in a low condition to allow for the absorption of flood discharges. River regulation also involves the reinforcement of minimum flows in the river channel downstream which obviously requires the maintenance of a large volume of water within the reservoir. Nevertheless, even if the reservoir storage is maintained at spillweir level, flood waves will be attenuated. The degree of attenuation will be related to a storage-head relationship and a head-outflow relationship. The former is controlled by the surface-area of the reservoir and is normally expressed as the retention factor (Lauterbach and Leder, 1969) which describes the attenuation of flood peaks through a reservoir as a function of the quotient of the total reservoir surface area to catchment area. The head-outflow relationship is controlled by the hydraulic characteristics of the spillweir - the greater the head maintained above spillweir level the greater the degree of outflow control. The detailed effects of a reservoir upon downstream discharges also depends upon the shape of the inflow hydrograph - short time-to-peak hydrographs will experience greatest attenuation - and the morphometry of the reservoir basin, which controls the rate of flood-wave passage.

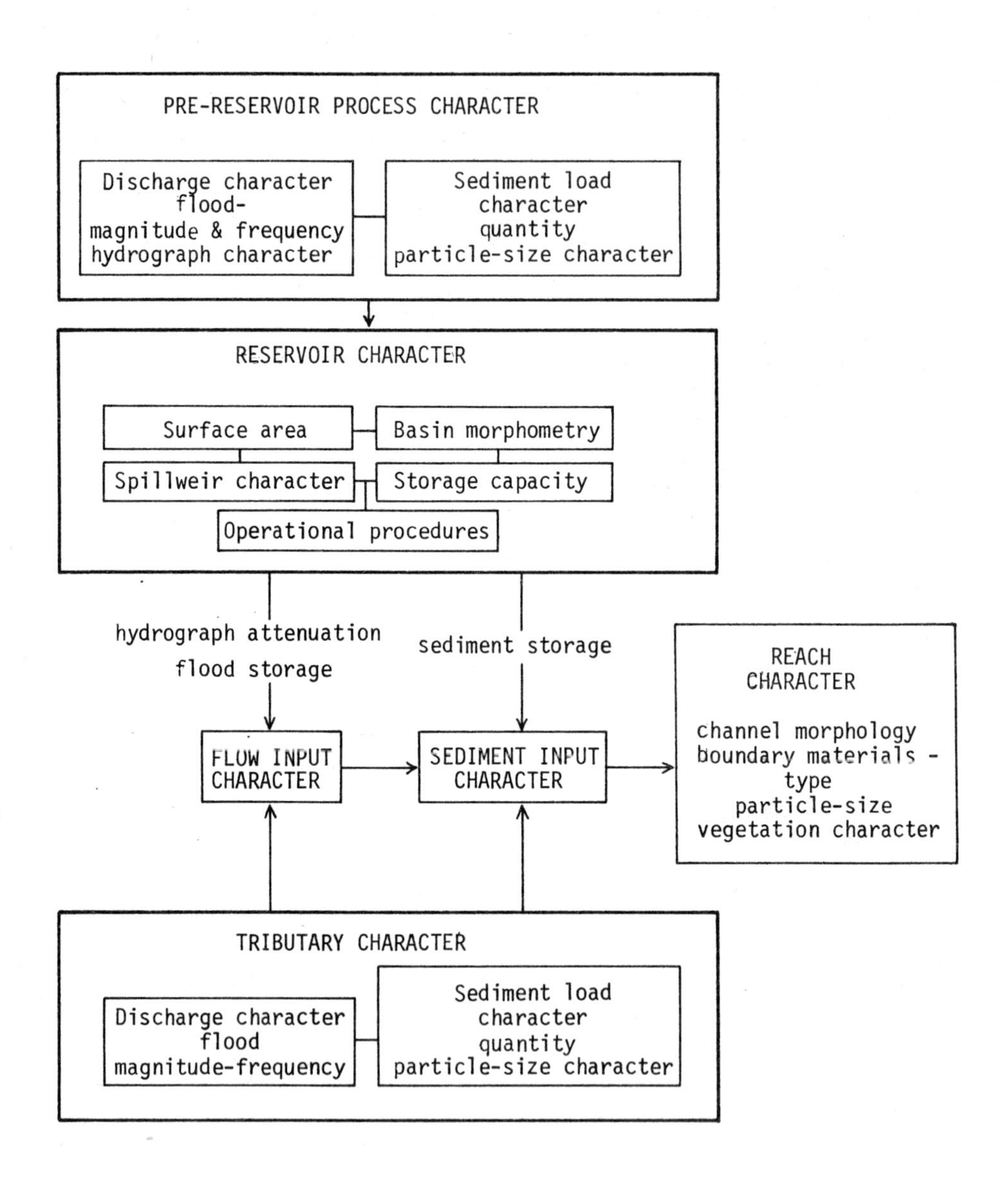

Figure 7.1. A framework for the examination of the effects of river impoundment upon channel morphology.

Whether the reservoir is full or empty, flood peaks will be reduced and the recession limbs of the hydrographs will be lengthened. Individual events may be completely eliminated by flood absorption or attenuated by temporary storage. Routing a flood through a reservoir with no free storage capacity may reduce peak discharges by up to 70 per cent (Petts and Lewin, 1978), but the effect of attenuation will decrease as the magnitude of the inflow hydrograph increases. The reduction of the magnitude of peak discharges may be greater for smaller, more frequent events and least for the larger, less frequent event because the attenuating effect of reservoir storage is independent of the size of the inflow events; low magnitude floods may be completely absorbed but the rare event may be insignificantly affected.

The reduction of flow velocities by lake storage results in the deposition of the sediment load transported by rivers feeding the reservoir. Part of the storage occurs in the reservoir itself but backwater effects induce the sedimentation of the coarser material in the channel and valley bottom upstream. Depending upon the velocity of the flow through the pool the fine material held in suspension will often be deposited in the reservoir proper but suspended sediment may on occasion be carried through the lake. The reservoir trap-efficiency, the percentage of incoming sediment trapped and deposited in a reservoir, depends upon a number of factors but the ratio of reservoir capacity to the average annual inflow, which describes the mean detention time of the stored runoff, has been shown to be the most important factor (Brune, 1953; Gotteschalk, 1964). Reservoirs act as artificial sediment sinks such that the sediment load delivered to the channel downstream is severely restricted and coarse debris is totally eliminated. Also, downstream from the dam the lag of the flood peak may desynchronise the mainstream and tributary discharges. Certainly the absorption of floods - particularly during summer when reservoir drawdown increases the available storage capacity - will lower the effective base-level for tributaries during storm events. Such changes could increase the sediment load from tributaries downstream from the dam. Thus, Makkaveyev (1970) observed that flow regulation by the Tsimlyansk dam in Russia lowered tributary base-level and induced tributary rejuvenation. In the first few years after dam completion, downcutting in the lower reaches of the Northern Donets increased tributary depths by three times and the discharge of sediment induced widespread aggradation within the channels of the lower Don. The sediment loads delivered to the mainstream by tributaries may also be altered by secondary impacts resulting from river impoundment. Dam construction in remote areas may provide a stimulus for other human activity such as forestry practices which can markedly influence channel changes below the impoundment. Increased sediment yields have been induced by road construction and logging practices (Beschta, 1978) and by drainage ditching associated with afforestation (Newson, 1980b).

Within an individual reach the fluvial processes will reflect the interaction of mainstream and tributary inputs operating on an existing channel form. Under natural conditions, a major storm event can result in one of two sediment output states depending upon the condition of the channel, itself a reflection of the relative recency, magnitude and character of preceding storm events, and upon the nature of the inputs into the reach. Indeed, variations of channel morphology are to be expected, reflecting changes in the erosional condition of the watershed prior to a storm event. A single storm may produce abnormally large sediment discharges so that the rate of supply into the channel reach may exceed the transport capability of the mainstream, resulting in net deposition. Alternatively, low sediment loads may be generated, the rate of supply will be less than the transport capability of the mainstream and the channel boundary materials may be eroded. In the long-term, erosional and depositional phases will form a balance maintaining morphological quasi-equilibrium. Channel adjustments downstream from dams will be governed by the interaction of two discharge frequencies in relation to the erodibility of the channel boundary and floodplain deposits. Both the rate and direction of channel change will be controlled by the interaction between the frequency of sediment loaded tributary events, effecting the construction of depositional forms within and along the mainstream and the frequency of competent reservoir releases which are responsible for the erosion of existing and introduced deposits.

3. CHANNEL CHANGES BELOW DAMS

In general terms a reservoir may induce one of three potential adjustments within each channel reach downstream (figure 7.2), depending upon the interaction of three parameters: the magnitude of flow regulation, the resistance of the boundary materials to erosion and the quantity and calibre of sediment delivered. Channels bounded by easily erodible materials having only a negligible effect upon flood magnitude and frequency may undergo erosion in response to the marked depletion of the sediment load. Indeed, channel degradation and scour has been repeatedly observed below dams as a result of clear water releases which abruptly increase a river's capacity to transport sediment. Commonly, channel depth and capacity will be increased, slope will be reduced and bed roughness may be increased so that, at least for floods of moderate frequency, the flow velocity will be reduced to below the threshold for sediment transport. Rashid (1979) for example demonstrated that the average channel bed degradation of the South Saskatchewan river below Gardiner Dam was 3.4 $cm.yr^{-1}$ and that the rate of thalweg degradation was five times greater than during the pre-dam period. However, the data provided reveals that 76 per cent of the four-year post-dam suspended load was transported in one year and perhaps 90 per cent of this was removed by one extreme summer flood,

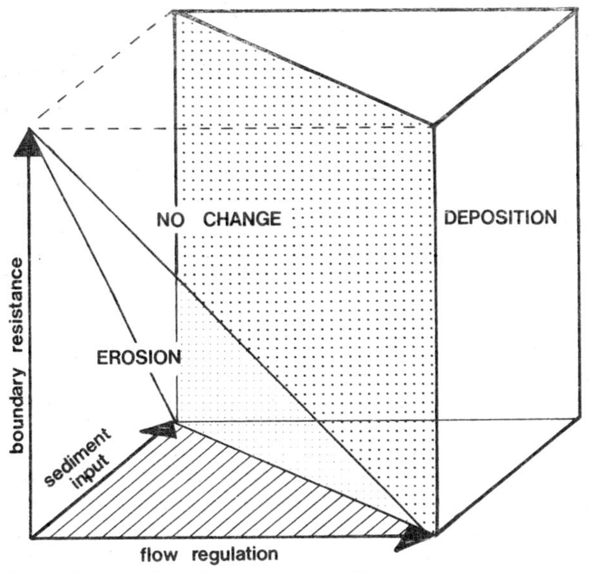

Figure 7.2. Channel adjustments in response to upstream impoundment (after Petts, 1980).

three years after project completion. In common with the majority of cases from which degradation has been reported, the bed of the South Saskatchewan is dominated by easily erodible, sand-sized sediments. Dam closure will not only increase river sediment transport capacity but will also diminish a river's competence to rework existing and subsequently introduced sediments. Thus, within gravel-bed rivers channel degradation may be improbable. Indeed, within the Green River below Flaming Gorge Dam, Utah, the reduction of peak flows has so limited the competence of the river that 93 per cent of the rapids are now stable as geomorphic/hydraulic features (Graf, 1980). Furthermore, if natural sediment loads are low, river impoundment will change the sediment loads for the river downstream relatively little, so that even if dams were built on or immediately above alluvial reaches, only limited degradation should be anticipated to occur, and at considerably slower rates than reported from heavily loaded rivers. The progressive realisation that degradation is not as problematical as formerly considered has led to an appreciation of the longer-term channel changes below dams which may be characterised by reduced channel capacities.

During the past 50 years sites of reduced channel dimensions have increasingly been reported below dams. Such a change of channel morphology requires the provision of a sediment supply and six sources have been identified (Petts, 1979) as windblown, upstream degradation, constructional practices, bank collapse, channel migration, and tributary injection. The redistribution of the channel boundary and floodplain sediments within actively migrating channels may prove particularly significant. Experimental releases from John Martin Dam on the Arkensas River resulted in channel scour which was characterised not by the longitudinal transportation of sediment but by local redistribution (Hathaway, 1948): the flows scoured the bed, in places by up to 60 cm and the material was moved laterally. The lateral migration of meanders together with the associated differential erosion and redistribution of floodplain deposits has reduced the channel capacity of the River Rede below Catcleugh Reservoir, UK, to twenty-five per cent of that expected under natural conditions. Nevertheless the major changes below dams have been induced by sediment deposition at and downstream from unimpounded tributaries.

In 1925 Salvador Arroyo wrote of fertile lowlands endangered by raising the bed of the Rio Grande. Deposition from outwash of small tributaries occurred as a result of mainstream flow regulation by the International-Elephant Butte storage and diversion scheme. On the Colorado River, prior to the construction of Glen Canyon Dam in 1963, deposits produced by tributary sediment discharges were periodically eroded by the frequent flows exceeding 1400 m^3s^{-1} (Dolan *et al*, 1974). Subsequently, the regulation of discharges has virtually eliminated flows of this magnitude. Indeed, at Lees Ferry, 24 km below the dam, the 10 year flood has been reduced from 2437 m^3s^{-1} to only 765 m^3s^{-1}. In 1972 and 1973 floods of the Little Colorado tributary produced mainstream bed aggradation of up to one metre and effected the lateral growth of bars and terraces by several metres. Photographic evidence for the period between 1923 and 1975 has revealed that deltas located at every major tributary confluence below Lees Ferry had become significantly enlarged, and that unstable canyon side-slopes had become stabilised by debris cones which were formerly eroded periodically by competent mainstream flows (Turner and Karpiscak, 1980).

The character of channel sedimentation

Sedimentation within regulated rivers occurs because the output of sediment from a mainstream is artificially slowed while the debris input from tributaries is, of course, unaffected or possibly increased. Consequent to dam construction the hydrological and sediment load characteristics are markedly altered and a phase of morphological disequilibrium will be induced. Sedimentation involving channel bed aggradation and bank deposition will occur at and downstream of tributary confluences because flows emerging from an uncontrolled tributary will spread across the wider mainstream reach in the absence of a simultaneous event, thereby decreasing its depth and velocity of flow. Initially the deposit will form a simple delta although the finer materials may be deposited along the channel margins downstream. According to Schumm and Hadley (1975) aggradation would continue until the bed of the channel becomes over-steepened, the deposit would then be trenched by headward erosion to form a channel that is once again in quasi-equilibrium with the discharge. The tributary's sediment load will then be transported through the reach and deposited in the next reach downstream and the cycle of

River	Reservoir	Date of completion	Channel capacity ratio of actual to estimated[1]		Primary adjustment
			above confluence	below confluence	
Shortclough	Leadhills	pre-1900	0.51	0.56	w^- d^-
Vyrnwy	Vyrnwy	1891	0.91	0.52	w^- d^-
Yeo	Blagdon	1900	1.03	0.58	w^-
Cow Gill	Cowgill	1904	0.94	0.51	w^-
Rede	Catcleugh	1905	0.60	0.31	w^- d^-
Camps Water	Camps	1925	0.85	0.60	w^-
Hodder	Stocks	1932	0.65	0.61	d^-
S. Teign	Fernworthy	1942	0.62	0.62	d^-
Derwent	Ladybower	1945	1.02	0.66	w^- d^-
Chew	Chew Valley	1953	0.90	0.60	w^- d^-
Yeo	Sutton Bingham	1955	0.70	0.47	w^- d^-
Avon	Avon	1958	0.97	0.57	d^-
W. Okement	Meldon	1970	1.01	0.53	w^- d^-

[1] The 'estimated' value in each case was determined from a regression analysis of channel dimensions and drainage area obtained from the field survey of upstream and neighbouring unimpounded channels.

(w = channel width, d = channel depth, - = decrease)

Figure 7.3. Some examples of channel change below tributary confluences along impounded British rivers.

River	Dam	Source	Observations on main channel	Observations on the role of vegetation
Logan Draw	Wyoming	King, 1961	3300 m^3 of aggradation at a tributary confluence after 2 years of mainstream flood control.	Stabilisation of sediments and further deposition induced by vegetation.
S. Canadian	Conchas	Hathaway, 1948	Aggradation at mouth of major tributary.	
Rio Grande	Elephant Butte scheme	Lawson, 1925	Formation of large fans at mouths of arroyos.	
		Arroyo, 1925	Channel bed aggradation.	
Republican	Trenton	Northrup, 1965	Channel cross-section reduced by up to 40%; sediment deposition associated with 10 major unimpounded tributaries.	Willows and cottonwoods stabilised bank materials and trapped up to 1 metre of sand in 4 years.
Peace	Bennett	Kellerhals and Gill, 1973	Substantial deltas observed below tributary confluences after 4 years of regulated flow.	Balsam poplars and willows stabilised gravel bars and deltas.
Sandstone Creek	Oklahoma	Bergman and Sullivan, 1963	Reduction of channel width.	Establishment of permanent riparian vegetation stabilised sediments.
Don	Tsimlyansk	Makkaveyev, 1970	Widespread aggradation associated with tributary rejuvenation.	

River	Dam	Source	Observations on main channel	Observations on the role of vegetation
S. Saskatchewan	Gardiner	Rashid, 1979	Channel width reduced from 220 m to 214 m.	
North Platte	Guernsey	Schumm, 1969	Channel narrowed from 1207 m to 46 m.	Trees became established on most of the floodplain.
Green River	Flaming Gorge	Graf, 1980	Increased stability of rapids and growth of rapids from tributary sediment sources.	
Colorado	Glen Canyon	Dolan et al, 1974	Expansion of debris fans by tributary washes.	Invasion of tamarisk and willow.
		Pemberton, 1975	Stabilisation of gravel-cobble bars has controlled channel degradation.	
		Turner and Karpiscak, 1980	Expansion of tributary deltas. Gullies along canyon sides no longer reactivated by removal of debris cones.	Stabilisation of tributary deposits by dense plant communities including salt cedar, arrowhead and seep willow which trap fines.
Bistrita	Izvoru Muntelui	Ichim and Radoane, 1980	Aggradation at 12 cm/yr^{-1} for 3 years after dam closure along 15 km reach.	

Figure 7.4. Observations on channel sedimentation below dams.

deposition and erosion will be repeated. However, observations along British rivers (Petts, 1979) revealed that changes of channel slope associated particularly with bar construction along the channel margin may serve to confine the flow, producing an increase in velocity and facilitating sediment transport through the reach.

Richards (1980) provided evidence to confirm that in floodplain channels the downstream variation of channel dimensions proceeds as a step function in which changes occur at tributary junctions, while within individual channel links the dimensions will vary stochastically about a mean in which little or no trend is apparent. Within a sample of thirteen impounded British rivers, however, the channel capacities below tributary confluences were relatively smaller, and often actually smaller than at locations upstream (figure 7.3). Indeed in many cases, for example below Bladgon, Meldon and Vyrnwy reservoirs, the morphology of channel sites above a tributary confluence corresponds to the estimated values for natural streams, but below the confluence channel capacities have been reduced to nearly one-half of those expected. In terms of channel shape, differences of channel response are related to the quantity and calibre of sediment introduced by a tributary in relation to the capacity and competence of the regulated mainstream flows: a major bed-material load producing channel bed deposits and finer materials producing a reduction of width by bank deposition and the construction of channel-side berms. Thus, discrete depositional and erosional phases separated by a threshold slope may occur simultaneously at different locations within a cross-section. By adjusting channel shape so as to increase the sediment transport capability of the section the progressive migration of the depositional front is facilitated. Nevertheless within each reach the relaxation path may be expected to approximate an exponential curve characterised by a progressively decreasing rate of change. King (1961), for example, observed the readjustment of the Logan Draw in Wyoming subsequent to the completion of an upstream flood-control project. Initially aggradation occurred rapidly at 1665 m^3yr^{-1} but this decreased to 185 m^3yr^{-1} after only seven years as the channel adjusted to the imposed discharge and sediment load conditions. The reaction time may, however, vary considerably for different locations along the channel; changes will be progressively induced downstream from the tributary confluence so that a considerable time-lag may exist before adjustments are initiated at locations remote from a tributary sediment source.

Recent studies of event effectiveness suggest, however, that the relaxation path will reflect not only the imposed frequency of mainstream flows and the frequency and magnitude of major sediment producing tributary events, but also the rate of colonisation and development of riparian and floodplain vegetation. Indeed, the reports of channel sedimentation at tributary confluences, or of channel width reduction in the absence of an immediate sediment supply (figure 7.4) have often emphasised the importance of vegetation both in stabilising deposits and in trapping fine materials. Prior to the construction of Harlan County dam, for example, extensive growths of willows and cottonwoods along the Republican River were prohibited by frequent abrasive mainstream events. Ten years after dam construction the active channel had been occupied by woody growth, which had trapped up to one metre of fine sand, and the length of active channel erosion had been reduced from 194 km to 32 km (Northrup, 1965). Thus, readjustment will depend not only upon the interaction of flows and sediment loads but also upon the climatic determinants of the growth of riparian floodplain vegetation. Variable rates and directions of actual channel response - as opposed to potential readjustment - will reflect different interactions between the frequency of competent reservoir releases, the frequency of sediment producing events within tributary sub-basins, and the rate of encroachment and growth of vegetation.

4. EVENT EFFECTIVENESS FOR CHANNEL CHANGE

Several authors have reported delta formation at tributary confluences along regulated rivers from a wide range of locations. Also, reports have described both the reduction of channel width and depth, and the conversion of divided channels into forms characterised by a well-defined course. However, a major conflict exists as to the importance of the deposits for channel adjustment. On the one hand it has been suggested that the deposits are only temporary storages and that the river channel will be maintained at its natural size or even enlarged by periodic mainstream events. This view is supported by studies of the magnitude of flow regulation produced by different reservoirs. The impact of a reservoir upon the magnitude and frequency of discharges downstream, as already observed, is related not only to the volume of storage space available for flood absorption, but also to the attenuation effect of the reservoir when storage is at or above spillweir capacity. Nevertheless, relatively small reservoirs inundating only a small proportion of their catchment and having only a limited capacity for flood absorption will have only a negligible effect upon high magnitude events whose potential for erosion within the channel downstream has been increased because of the sediment load abstraction. An alternative view is that the modification of the channel dimensions and energy slope achieved by depositional activity will prohibit erosion and facilitate a readjustment of the channel form to a new quasi-equilibrium associated with the imposed flow regime and sediment load characteristics. Indeed, studies of the frequency of sediment transporting events have demonstrated that the amount of work which may be achieved by flows below dams, in terms of annual sediment transport may be considerably reduced. Bondurant and Livesey (1973), from experience of reservoir operation in the United States, concluded that the total capacity for sediment transport of an impounded river may be reduced by up to 75% depending upon the relative characteristics of the controlled release versus the normal flow. For a variety of reasons studies of the relaxation paths for channel forms affected by upstream impoundment have proved problematic, not least because of the often long-time period involved. Nevertheless, consideration of the channel changes as the integration of the effects of destructional mainstream flows, constructional tributary discharges and the factors influencing channel stability may prove illuminating. First, an example of each of the above contentions will be discussed.

Nearly fifty years after dam construction the channel of the Tujunga Creek (figure 7.5A) has failed to achieve a stable form adjusted to the imposed process conditions. The Tujunga Creek drains a 300 km^2 catchment within the San Gabriel Mountains of Southern California. The basin has a semi-arid climate characterised by a marked summer drought, although winter rains often exceed 500 mm. The underlying rocks are relatively resistant to erosion but the intense fracturing, high relief, and steep slopes, together with a sparse vegetation cover, combine to produce high sediment yields. Reservoir sedimentation rates indicate a long-term mean annual sediment yield of 735 $m^3 km^2 yr^{-1}$ but this has

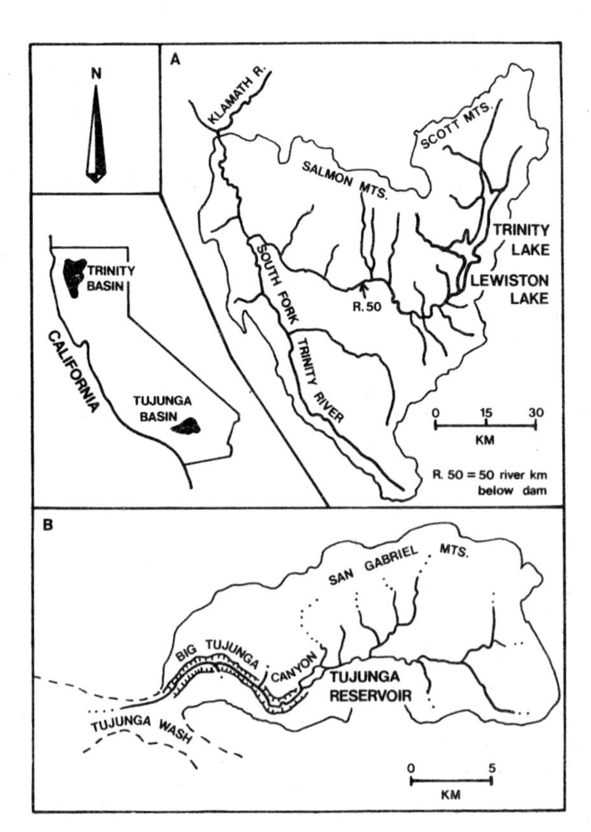

Figure 7.5. Location map for the Tujunga and Trinity river basins.

been exceeded during the period 1966-1976. The ten-year period produced an increased mean annual sediment yield of 1925 $m^3 km^2 yr^{-1}$ which may reflect the increased incidence of fires, often suspected to have been man-induced, and other human activities within the watershed such as road construction, which were of a relatively high intensity during this period.

The Tujunga Creek system may be viewed as composed of two zones: an erosional source area and a depositional zone where the creek enters the fan-head valley of the Tujunga Wash. The source area is characterised by geologically recent uplift and continuing tectonic activity, and the creek has become confined within the Big Tujunga Canyon which is fed by numerous steep ephemeral tributaries. Tujunga Wash is a 4 km^2 depositional zone which has formed as a result of a change of the hydraulic factors which produce a reduction of flow competence as the creek leaves the confines of the canyon. However, fifteen kilometers upstream from the Wash the transfer of sediment is interrupted by the Big Tujunga Dam. Indeed the dam completed in 1931 impounds an area of 211 km^2, some seventy per cent of the source area. The reservoir is relatively small inundating only one-third of one per cent of the catchment and its storage capacity has already been reduced by 70 per cent to only 3000 acre-feet as a result of high sediment yields. Thus, major flood discharges pass the spillweir only slightly attenuated although frequent, low magnitude events may be controlled.

The negligible effects of the reservoir upon flood discharges combined with the isolation of a large source area for sediments may be expected to produce conditions favourable to rapid degradation below the dam. Forty-nine years after project completion a stable, clearly defined channel has developed at only a few isolated locations. Within most of the 15 km reach between the dam and the Wash the channel banks are poorly defined and at many locations the entire canyon floor becomes active during flood discharges. Massive fans have developed at the mouths of ephemeral tributaries constricting channel width, changing the local channel slope and diverting the mainstream channel course into the opposite bank inducing localised erosion. Although the active floodplain width has been reduced locally the character of the channel demonstrates no clearly defined response to headwater impoundment. The sediment load has been efficiently trapped by the dam but extensive aggradation of up to three metres has been observed at the mouth of the canyon during extreme storm events (Scott, 1973). This reflects the removal of sediment accumulations from the canyon floor below the dam which were derived from periodic tributary injection. A long-term balance between the regulated flows and channel morphology has been prevented by the existence of conflicting destructional and constructional processes which are both highly effective, at least in the short-term. Channel sedimentation is produced by frequent tributary discharges but in the absence of a well-established vegetation cover the less frequent mainstream flows serve to flush much of the tributary injected debris downstream.

In contrast, a relatively rapid adjustment to a new stable, quasi-equilibrium form has been attained along the Trinity River. The United States Bureau of Reclamation completed construction of the Trinity River Project in the early 1960s to provide a major source of hydro-electric power, and for the diversion of water to the Sacramento system. The river originates in the Scott Mountains of Northern California and at its confluence with the Klamath River, 65 km from the Pacific, has a catchment of 7637 km^2 (figure 7.5B). Mean annual rainfall within the 1885 km^2 reservoir catchment approaches 1200 mm with a marked seasonal variation, producing high discharges in winter and spring but only nominal flows in summer. The mountainous catchment supports a 90% coniferous forest cover but sediment loads are moderately high. However, the large capacity reservoir, which inundates 3.67% of the impounded basin, is affected to only a minor extent by sedimentation. Prior to dam construction in 1963 the Trinity River below the dam site was capable of transporting an average of 150 000 m^3 of bedload per year (D.W.R., 1978). Temporary deltas were formed at the mouths of tributaries but the deposits were regularly flushed downstream by the frequent competent mainstream flows. Indeed, under historical conditions at Lewiston, flows of sufficient magnitude to transport riffle gravels were generally exceeded between 10 and 40% of the time (D.W.R. pers. comm.). For 50 km downstream of Lewiston the channel actively migrated across a sand and gravel floodplain which supported only a sparse vegetation cover. Through this reach the Trinity River channel ranged from 35 to 100 m wide with a mean width of 65 m.

Subsequent to project completion the Trinity River downstream of the dams has been significantly altered. The headwater sediment sources have been isolated from the river below the dam but it is the discharge regulation provided by Clair Engle Lake that has had the dominant effect. The magnitude of the post-dam ten-year discharge is only one third of that described by the pre-reservoir data and the 1.5 year flood has been reduced from 350 m^3s^{-1} to less than 15 m^3s^{-1}. The duration of the minimum discharge to move gravels has been reduced from 40% of the time to less than three per cent, and this has resulted in a reduction of the sediment transport capacity to approximately 8000 m^3 - less than one-quarter of the sediment load delivered annually by one tributary, Grass Valley Creek (D.W.R., 1978). Channel sedimentation has been induced, deltas composed of flood gravels introduced by the flash flooding tributaries have expanded, pools have filled with fines and channel side berms, formed mainly of sand, have been constructed downstream from the tributary confluences.

Examination of aerial photographs has revealed that channel widths have been reduced by up to 65% for 50 km below the Lewiston Dam (figure 7.6). The duration of competent discharges increases only slowly downstream to the confluence with the North Fork and this is reflected by the channel morphology. Ten years after project completion channel widths ranged between 25 and 65 m with a mean of 33 m - a reduction of over 30 m when compared with the pre-

Channel length below dam (km)	Drainage area impounded (%)	Percentage of time minimum discharge required to move gravels equalled or exceeded[1] Pre-dam	Post-dam	Channel width reduction (%)
0 -	99	50	0.05	0 - 30
4 -	93	45	0.05	0 - 60
9 -	88	45	3	50 - 60
20 -	83	25	1	30 - 40
23 -	78	40	3	30 - 55
25 -	76	40	5	33 - 65
31 -	70	35	5	no data
43 -	64	30	4	20 - 38
53	51	60	40	no data

[1] Data provided by Department of Water Resources, Red Bluff, California.

Figure 7.6. Channel adjustment below Trinity and Lewiston Dam, Trinity River, North California.

reservoir values. However, the detailed examination of a four hundred metre reach (figure 7.7), incorporating the Grass Valley Creek confluence, reveals that the major adjustments of channel morphology occurred during the ten year period immediately subsequent to project completion and that only limited readjustments have occurred subsequently. Between 1962 and 1972 the area of flood plain increased by 14% within this section at the expense of the Trinity River channel. Moreover, the vegetation density increased from a cover of 58% to over 90%. During the subsequent five years the floodplain area has been enlarged only to a minor degree but vegetation development has continued, to provide a 98% cover primarily of willows and alders in 1977. Channel changes downstream from Lewiston Dam have occurred rapidly, facilitated by a high degree of flow regulation, a high frequency of sediment-loaded tributary discharges in relation to the frequency of sediment depleted mainstream flows, and adequate environmental conditions for the rapid establishment and growth of riparian and floodplain vegetation. Even a 'major' flood in January 1974 which produced competent discharges for eleven consecutive days failed to affect the riparian vegetation and had little impact with respect to moving gravels and cobbles from the deposits. Channel width has been reduced, the velocity and depth components of the at-a-station hydraulic geometry have increased and many reaches appear to be in quasi-equilibrium with the contemporary flow characteristics and tributary sediment loads. It is reasonable to propose that whilst channel readjustment

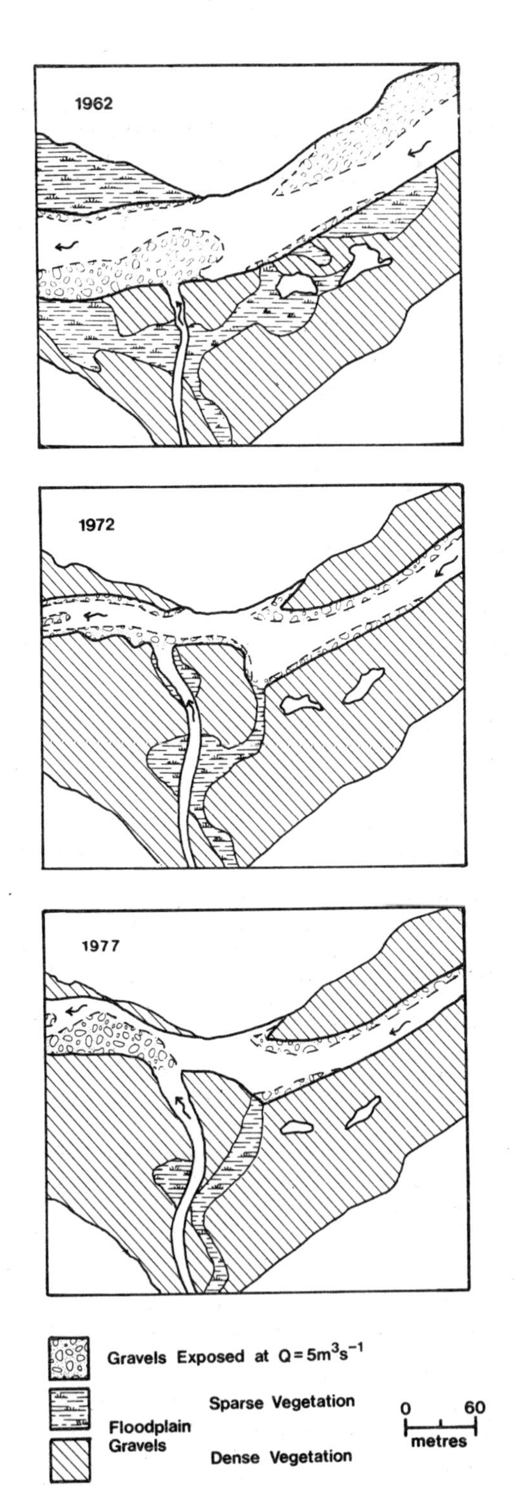

Figure 7.7. Channel changes at the Grass Valley Creek - Trinity River confluence since mainstream impoundment in 1963.

may have resulted from the extreme conditions of flow regulation alone, vegetation establishment certainly reduced the relaxation time by protecting the deposits from occasional erosion, and may also have increased the degree of channelisation which has exaggerated the reduction of channel width.

The relaxation path

Attempts to develop a general model of channel adjustment are problematic, not least because of the lack of long-term repeated observations and because of the complex variability of channel response in the short-term. Nevertheless channel response to upstream impoundment may be viewed as a relatively simple system because the behaviour of each control variable is for most situations uni-directional: flood-magnitudes and sediment loads will be more or less reduced, tributary inputs will increase in significance and the effect of riparian vegetation may also be increased. It is therefore possible to produce a series of hypothetical models to describe the complex response of channel morphology when subjected to particular conditions. The characteristic relaxation paths produced will reflect the relative frequencies of the competent reservoir releases and the sediment-loaded tributary flows.

Downstream from reservoirs having only a negligible impact upon the frequency distribution of competent discharges an erosional response may be induced as a result of the isolation of sediment sources (figure 7.8A). Such a response may be induced by either relatively small reservoirs having an insignificant effect upon flood discharges yet effectively trapping the sediment load or by the existence of sand-sized channel bed sediments which may be easily eroded even by regulated flows. Under such conditions the relaxation time will reflect the frequency of competent discharges and the reaction time will reflect the actual time period between dam closure and the occurrence of a competent reservoir release. Gravel bed rivers require rare, high magnitude events to induce sediment movement so that a considerable time lag may exist before a morphological change is induced: within sand-bed channels response may be immediate. Armouring may constrain channel response, but generally the channel dimensions will be progressively enlarged until the hydraulic conditions are so modified as to prohibit further erosion. However, within heterogeneous bed-materials the differential transport of particular size-fractions may produce an incomplete but protective layer of relatively coarse particles prohibiting further bed scour.

Although an erosional potential may exist within a regulated river the channel capacity may be maintained at its pre-reservoir dimensions (figure 7.8B). The accumulation of sediments, derived from non-regulated tributary sources, within the main channel during the intervals between competent reservoir releases may satisfy the capacity of these releases and prohibit channel enlargement over time. Nevertheless, considerable short-term variations may be observed which reflect the actual time period between depositional tributary events and erosional mainstream discharges. In contrast, below large flood-control reservoirs having a major impact upon the frequency of competent flows a potential for channel readjustment to reduced dimensions may dominate. In the absence of a sediment supply, however, the flows will be accommodated within the existing channel form. For channels with finer bed sediments gradual erosion may actually occur, induced by rare competent discharges depleted of sediment load. Such a readjustment is described by figure 7.8A, although the reaction and relaxation times may be considerably increased. Alternatively, aggradation resulting from the injection of sediment by tributaries may be opposed and, in the long-term, prevented by infrequent flood discharges passing the dam and effectively eroding the deposits (figure 7.8C). Considerable variations of channel form would be observed at different points in time but in the long-term the natural channel dimensions may be maintained. In an extreme case (figure 7.8D) the complete removal of competent mainstream discharges would result in the progressive deposition of tributary injected sediments until the channel dimensions had readjusted so as to provide the velocities required for the transmission of the sediment loads.

The effectiveness of reservoir releases depends, however, not only on the particle-size composition of the deposits and volume of sediment injected, but also upon the rate of establishment and growth of vegetation. Under natural conditions the extensive growth of floodplain and riparian vegetation is often prohibited by relatively frequent flood flows which disturb floodplain deposits and have an abrasive effect upon seedlings and perhaps larger plants. The reduction in frequency of flood discharges and the provision of stable low flows may encourage vegetation encroachment which would stabilise deposits, trap further sediments and reduce the length of active channel erosion. The examination of regulated rivers has suggested that both sediment injection from unimpounded tributaries and sediment stabilisation by vegetation play an important role in channel adjustment. The response of channel morphology downstream from a major sediment source within a regulated river which still experiences destructive mainstream flows of moderate frequency (eg. figure 7.8C) may be described in more detail by inclusion of the time-period required for vegetation establishment on the new deposits (figure 7.9). The reaction time (L1) will simply reflect the time period between dam closure and the first tributary event but the relaxation time (L2) will be dependent primarily on the availability of conditions favourable to vegetation growth. Impoundment may create a potential for a reduction of channel capacity but prior to vegetation establishment mainstream events (D1 and D2) will erode the deposits introduced by tributary flows producing temporary enlargements of the channel dimensions. However, the actual time period between mainstream events during 'A' is sufficient to permit vegetation establishment so that the effectiveness of subsequent mainstream events may be considerably reduced (D3). Once a

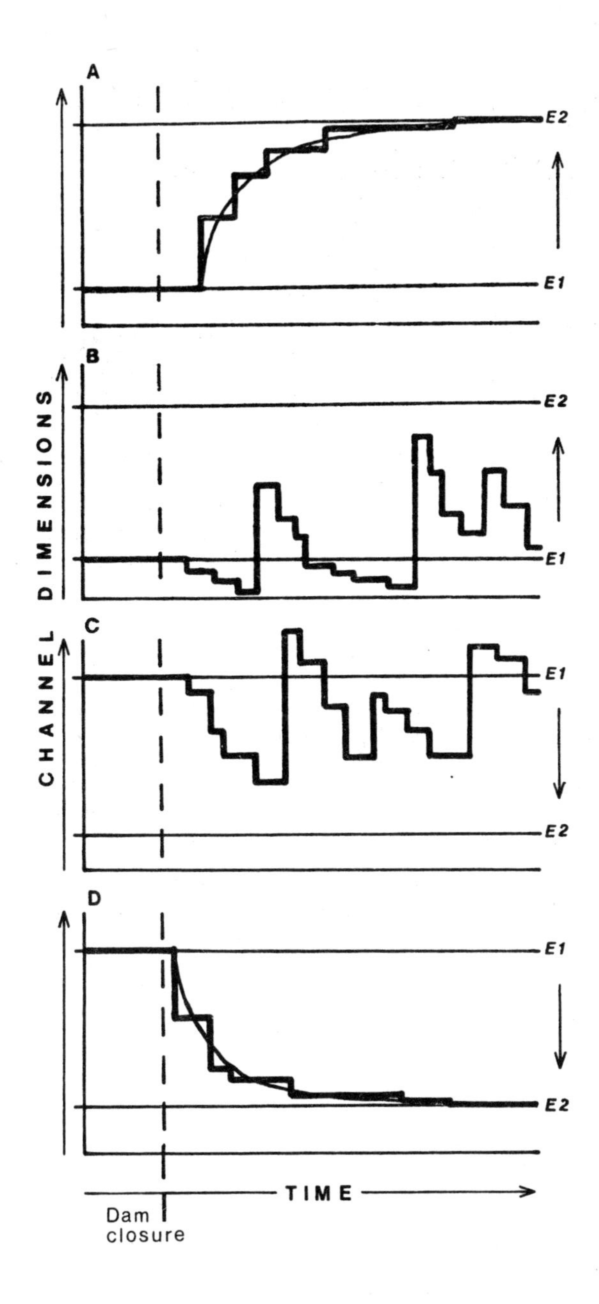

Figure 7.8. Hypothetical responses of channel morphology to headwater impoundment. Potential responses involving a change from an initial equilibrium state (E1) to a potential equilibrium state (E2) are viewed as a continuum with two clearly defined extremes characterised by degradation and scour (A) and aggradation (D).

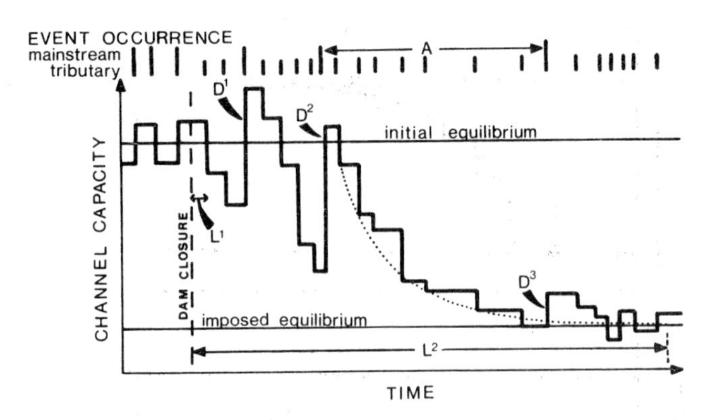

Figure 7.9. Hypothetical response of channel morphology within a regulated river below a major tributary sediment source. The relaxation path reflects the frequency of major reservoir releases (D1, 2, 3) modified by tributary sediment injection and vegetation establishment during 'A'.

strong vegetation cover has developed relatively minor fluctuations of channel form will occur in response to varying discharge-sediment load combinations supplied by the sub-basins. The readjustment of the channel dimensions will continue until adequate velocities are provided to prevent net deposition and a new quasi-equilibrium condition will be attained. Thus, variable rates and directions of actual channel response - as opposed to potential adjustment - will reflect different interactions between the frequency of competent reservoir releases, the frequency of sediment producing events within tributary sub-basins, and the rate of encroachment and growth of vegetation. For individual locations the relaxation time may be relatively short but because of the sequential nature of sediment transport the complete readjustment of the channel form downstream from a tributary confluence may require a long period of time.

5. CONCLUSIONS

River channel readjustments induced by human impacts have increasingly proved a source of attraction for researchers during the past two decades. The majority of studies have, however, adopted a 'case-study' approach. Few attempts have been made to interpret the spatially variable channel responses in terms of discharge and sediment load changes in relation to the growth of riparian and floodplain vegetation. Knox (1972) suggested that episodic phases of landform adjustment in response to climatic change are markedly influenced by the rate of vegetation establishment and growth. Sudden changes of fluvial processes as a result of human activity may produce comparable situations. Wolman and Gerson (1978) defined process effectiveness as the ability of an event or combination of events to affect the shape or form the landscape. However, to date an effectiveness approach has been primarily directed to the examination of the destructive events (e.g. Newson, 1980a) under natural conditions. The evidence presented here suggests that an effectiveness approach which directs attention to the complete interaction of processes, forms, and vegetation may provide for an improved understanding of the complex responses of river channels to human impacts.

Dams may induce a continuum of channel changes downstream and the two extreme conditions may be easily identified. In either case, within a particular reach channel change will initially be rapid and the rate of change will progressively decrease. Downstream from relatively small reservoirs, within sand bed channels or in environments where the rate of vegetation establishment is slow and growth minimal, mainstream events should produce progressive degradation. In contrast, below large flood-control reservoirs, within gravel-bed rivers, or in locations of rapid vegetation establishment, and assuming a supply of sediment a relatively rapid readjustment may be achieved, characterised by a reduction of channel width and of channel capacity. Indeed, providing that a sediment supply is available under conditions of flow regulation, the potential readjustment of channel morphology may be characterised by a trend towards reduced cross-sectional dimensions. Sedimentation within regulated rivers will only be a temporary condition, however, if the frequency of competent mainstream releases prohibits vegetation establishment.

Failure to accept that different morphological responses can be induced by application of the same stimulus has been a major stumbling block in public perception which, once overcome, will provide an improved understanding of river management problems. Sedimentation within regulated rivers is increasingly being reported as a cause of serious ecological and economic concern. Although depositional activity within channels downstream from dams has been observed for more than fifty years, research has been dominated by investigations of channel degradation and scour. Indeed, the potential severity of mainstream aggradation induced by headwater impoundment has often been neglected in impact assessments during project formulation. The significance of induced sedimentation for the channel form in the long-term, however, will be related to the interaction of three factors: the frequency of competent mainstream flows and the capacity of those flows, the frequency of sediment producing events from tributary subbasins and the calibre of the materials derived, and the rate of vegetation encroachment and growth, which is related in part to climatic determinants but also to the frequency of inundation and substrate activation. Thus, the application of an effectiveness approach to man-induced channel changes, directed both at the effectiveness of destructional and constructional processes, may provide for improved interpretations of the spatial complexity of channel response and for the development of improved models to predict the long-term consequences of human activity.

ACKNOWLEDGEMENTS

Special thanks are due to E.F. Serr and his colleagues at the Department of Water Resources, Red Bluff, California, and to Professor A. Orme and his assistants and students at the University of California in Los Angeles, for making field work possible, for providing data and for their enthusiastic discussions. I would also like to thank Dr. M. Newson for his constructive comments during the preparation of this paper. Thanks are also owed to Mrs J. Jarvis and to Miss H. Briars for patiently typing the manuscript and preparing the figures.

REFERENCES

1. Allen, J.R.L., 1974, Reaction, relaxation and lag in the natural sedimentary systems: general principles, examples and lessons, *Earth Science Reviews, 10*, 263-342.

2. Allen, J.R.L. 1977, Changeable rivers: mechanisms and sedimentation, in: *River Channel Changes*, ed. Gregory, K.J., (Wiley, Chichester), 15-74.

3. Arroyo, S., 1925, Channel improvements of Rio Grande below El Paso, *Engineering News Record, 95*, 374-376.

4. Begin, Z.B., Meyer, D.F. and Schumm, S.A., 1981, Development of longitudinal profiles of alluvial channels in response to base level lowering, *Earth Surface Processes, 6(1)*, 49-68.

5. Bergman, D.L. and Sullivan, C.W., 1963, Channel changes on Sandstone Creek near Cheyenne, Oklahoma, *United States Geological Survey, Professional Paper 475c*, 145-148.

6. Beschta, R.L., 1978, Long-term patterns of sediment production following road construction and logging in the Oregon Coast Range, *Water Resources Research, 14(6)*, 1011-1016

7. Bondurant, D.C. and Livesey, R.H., 1973, Reservoir sedimentation studies, in: *Man Made Lakes*, ed. Ackerman, W.C., White, F.G., Worthington, E.B., (American Geophysical Union), 364-367.

8. Brune, G.M., 1953, Trap efficiency of reservoirs, *Transactions of the American Geophysical Union, 34*, 407-418.

9. Dolan, R., Howard, A. and Gallenson, G., 1974, Man's impact on the Colorado River in the Grand Canyon, *American Scientist, 62*, 392-401.

10. D.W.R., 1978, Grass Valley Creek: sediment control study, *California Department of Water Resources Report*, Northern District, Red Bluff, 74 pp.

11. Gotteschalk, L.C., 1964, Reservoir sedimentation, in: *Handbook of Applied Hydrology,* ed. Chow, V.T., (McGraw-Hill, New York), section 17-1.

12. Graf, W.L., 1977, The rate law in geomorphology, *American Journal of Science, 272*, 178-191.

13. Graf, W.L., 1979, The development of montane arroyos and gullies, *Earth Surface Processes, 4*, 1-14.

14. Graf, W.L., 1980, The effect of dam closure on downstream rapids, *Water Resources Research, 16(1)*, 129-136.

15. Hathaway, G.A., 1948, Observations on channel changes, degradation and scour below dams, *International Association of Hydraulic Research,* Second Congress, Stockholm, 267-317.

16. Ichim, I. and Radoane, M., 1980, On the anthropic influence time in morphogenesis, with special regard to the problem of channel dynamics, *Revuew Roumaine de Geographie, 24*, 35-40.

17. Kellerhals, R. and Gill, D., 1973, Observations and potential downstream effects of large storage projects in Northern Canada, *Proceedings of the eleventh Congress of the International Commission on Large Dams,* Madrid, Spain, 731-754.

18. King, N.J., 1961, An example of channel aggradation induced by flood control, *United States Geological Survey, Professional Paper 424B,* 29-32.

19. Knox, J.C., 1972, Valley alluviation in south-western Wisconsin, *Annals of the Association of American Geographers, 62*, 401-410.

20. Lauterbach, D. and Leder, A., 1969, The influence of reservoir storage on statistical peak flows, in: *Floods and their computation, 2*, International Association of Scientific Hydrology, 281.

21. Lawson, J.M., 1925, Effects of Rio Grande storage on river erosion and deposition, *Engineering News Record, 95(10),* 372-374.

22. Makkaveyev, N.I., 1970, The impact of large water engineering projects on geomorphic processes in stream valleys, *Geomorphologia, 2*, 28-34.

23. Newson, M., 1980a, The geomorphological effectiveness of floods - contribution stimulated by two recent events in mid-Wales, *Earth Surface Processes, 5*, 1-16.

24. Newson, M., 1980b, The erosion of drainage ditches and its effect on bed-load yields in mid-Wales: reconnaissance case studies, *Earth Surface Processes, 5(3),* 275-290.

25. Northrup, W.L., 1965, Republican river channel deterioration, *United States Department of Agriculture Miscellaneous Publication, 970*, 409-424.

26. Pemberton, E.L., 1975, Channel changes in the Colorado River below Glen Canyon Dam, *Proceedings of the 3rd Federal Interagency Sedimentation Conference,* Water Research Council, Washington, 5.61-5.73.

27. Petts, G.E., 1979, Complex responses of river channel morphology subsequent to reservoir construction, *Progress in Physical Geography, 3(3),* 329-362.

28. Petts, G.E., 1980, Implications of the fluvial process-channel morphology interaction between British reservoirs for stream habitats, *The Science of the Total Environment, 16,* 149-163.

29. Petts, G.E. and Lewin, J., 1979, Physical effects of reservoirs on river systems, in: *Man's impact on the hydrological cycle in the United Kingdom,* ed. Hollis, G.E., (Geo Abstracts, Norwich), 79-92.

30. Rashid, H., 1979, The effects of regime regulation by the Gardiner Dam on downstream geomorphic processes in the South Saskatchewan river, *Canadian Geographer, 23,* 140-158.

31. Richards, K.S., 1980, A note on changes in channel geometry at tributary junctions, *Water Resources Research, 16(1),* 241-244.

32. Schumm, S.A., 1969, River metamorphosis, *Journal of the Hydraulics Division, Proceedings of the American Society of Civil Engineers, 441*, 255-273.

33. Schumm, S.A., 1977, *The Fluvial System,* (Wiley, New York).

34. Schumm, S.A. and Hadley, R.F., 1957, Arroyos and the semi-arid cycle of erosion, *American Journal of Science, 255*, 164-174.

35. Scott, G.R., 1973, Scour and fill in Tujunga Wash - a fanhead valley in urban south California, *United States Geological Survey, Professional Paper 732B.*

36. Turner, R.M. and Karpiscak, M.M., 1980, Recent vegetation changes along the Colorado River between Glen Canyon Dam and Lake Mead, Arizona, *United States Geological Survey, Professional Paper 1132.*

37. Wolman, M.G., 1967, Two problems involving channel changes and background observations, *Quantitative Geography,* Part II, Northwestern University Studies in Geography, *14*, 67-107.

38. Wolman, M.G. and Gerson, R., 1978, Relative scales of time and effectiveness in watershed geomorphology, *Earth Surface Processes, 3*, 189-208.

39. Wolman, M.G. and Miller, J.P., 1960, Magnitude and frequency of forces in geomorphic processes, *Journal of Geology, 68*, 54-74.

8. The cover sands of north Lincolnshire and the Vale of York

P.C. Buckland

INTRODUCTION

During the mapping of Lincolnshire and Yorkshire by the officers of the Geological Survey in the late nineteenth century, extensive spreads of 'blown sand' were recorded along the eastern side of the Vale of York, from the neighbourhood of Northallerton, southwards into the Trent Valley, at least as far south as Collingham and eastwards to the foot of the Lincolnshire Wolds. The term 'blown sand' was employed for superficial sands of varying age and origin, ranging down to the still active dunes of parts of Humberside. Apart from some recorded sections in the sheet memoirs (Dakyns *et al.*, 1886; Ussher, 1890), these deposits gained little comment. Wilson (1948) only refers to post-Glacial blown sand and the most recent relevant sheet memoir, for the Ollerton district (Edwards, 1967), has a similar comment. Edwards (1936) noted dreikanter in the Pleistocene succession in the Vale of York and, although most of his localities suggested an origin prior to the maximum ice advance at the last glaciation, Kendall and Wroot (1924) had noted a similar occurrence of wind faceted boulders over glacial deposits north of the York-Escrick moraine complex, suggesting more than one period of aeolian activity. Swinnerton (1914) had also obtained evidence for Late Pleistocene dreikanter formation in south Nottinghamshire.

Few data on the blown sand were forthcoming from geological sources, however, and the first dating evidence was provided, inadvertently, by archaeologists working in north Lincolnshire and south Humberside. Armstrong (1931) published an assemblage of flint artifacts from a site in the blown sand at Sheffield's Hill, 5km north of Scunthorpe. Despite the associated forest vertebrate fauna, he regarded the site as an Upper Aurignacian (Creswellian) hunting station and later excavated two similar sites in the sands near Willoughton, 11km east of Gainsborough (Armstrong, 1932a and b). Although subsequent research has shown that these assemblages, with backed blades, obliquely blunted points, bipolar cores, end scrapers and a few smaller geometric implements, belong to the mesolithic (Radley, 1969; Buckland & Dolby, 1973), Armstrong's discussion did imply that the blown sand, in part, belonged to either the Devensian or early Flandrian. Armstrong (1932b; 1956) was also familiar with other flint artifact finds from the blown sand areas of south Humberside and saw, in the small geometric microliths, a relationship with the Tardenoisian of France and the Low Countries. The abundant finds from Risby Warren, 4km northeast of Scunthorpe, formed, in his opinion, the type for this cultural grouping in Britain. Whilst its European connections must now be treated with more circumspection (Mellars, 1974), this material belongs to the later mesolithic and, since the finds are incorporated in the sands, suggests that these had been emplaced by this period. Extensive archaeological fieldwork by Dudley (1931; 1949) and others supports this but the still active nature of dunes on some of the warrens around Scunthorpe makes it difficult to obtain secure stratigraphic relationships. Two flint implements from Risby Warren, however, were found in a clay layer at the base of the blown sand. These were a Mousterian hand-axe and a typically Upper Palaeolithic shouldered point (Lacaille, 1946; May, 1976), suggesting that the sands were deposited in the latter half of the Devensian. Several other typologically Late Upper Palaeolithic (Creswellian) pieces are known from Risby (May, 1976). Dudley (1949) also recorded teeth and a tusk of mammoth from gravel pockets beneath the blanket of sand at Flixborough and Thealby, north of Scunthorpe. Neither occurrence, however, provides anything more than a broad *terminus post quem*.

When Straw (1963) examined section in the sands around Caistor, on the scarp slope of the Lincolnshire Wolds, and at Crosby Warren, near Scunthorpe, he found stratigraphic evidence for the influence of periglacial conditions and suggested on geomorphological grounds that the sands were in part contemporary with the last advance of the Devensian ice sheet and in part subsequent to its retreat from the foot of the Chalk dipslope. Two ^{14}C dates from Dimlington, on the Yorkshire Coast, imply that the maximum ice advance post-dates c. 18 500 B.P. (Penny *et al.*, 1969). Straw further suggested that the sands were emplaced before the establishment of a complete vegetation cover in the post-Glacial and he tentatively ascribed the podzolisation of the deposits to the Atlantic period. Deposition must therefore have ceased no later than the Late Boreal (c. 8000 B.P.). He recognised several possible origins for the blown sands and introduced the non-committal term 'Cover Sands', which was already widely used for similar deposits in the Low Countries (Maarleveld, 1960).

At Stockton-on-the Forest, north-east of York (SE 608640), Matthews (1970) obtained a ^{14}C date of 10 700 ± 190 B.P. (N-488) on a layer of compact, humified peat at a depth of 1.5m in the Cover Sands, 0.2m above a till. Nearby, a further ^{14}C date of 9950 ± 180 B.P. (N-820) was gained from a frost contorted layer of gyttja, 1.3m above the base of the Sands. At least some of the Cover Sands are therefore dated to the terminal cold phase of the last glaciation, the Loch Lomond Stadial (Younger Dryas), although the small amount of sand below the first dated horizon, partially filling frost wedge casts, could belong to any time from the retreat of the Devensian ice sheet from the York Moraine to the first part of the Stadial. Particle size analysis confirmed the aeolian nature of the deposits and a pollen count from close to the base of the second section implied a Pollen Zone I or III age. Gaunt, Jarvis and Matthews (1971) examined the Late Devensian sequence in the Vale of York on a regional basis and placed this aeolian phase between an older fluvial depositional phase, associated with the drainage of pre-glacial Lake Humber, and a younger phase of aggradation from a base level about 15m below the present. A buried soil on top of the older deposits (the '25 foot Drift') at Armthorpe (SE 649072) was dated to 11 100 ± 200 B.P. (N-810). At Cawood, near Selby, Jones and Gaunt (1976) obtained a date of 10 469 ± 60 B.P. (SRR-870) and a pollen assemblage from peat at the base of the sands, and Gaunt (1981) has recently provided a useful summary of the Quaternary geology of the southern part of the Vale of York. The age of some of the Cover Sands was thus fixed at three widely spaced localities as lying within the terminal cold phase of the Devensian. The final cessation of sand blowing, however, apart from post-forest clearance movement, has yet to be ascertained.

STRATIGRAPHY

Mineral extraction in South Humberside provides extensive exposures in the Cover Sands. Around Messingham and the west side of the Isle of Axholme, near Haxey, the sand is quarried for use in glass-making and for refractories. North and east of Scunthorpe, the mining of the Frodingham Ironstone provides many ephemeral exposures and on both Hatfield Chase and the Trent floodplain, motorway construction (M18, M180) and drainage ditches have also given useful sections. In the smaller exposures, however, it is often difficult to define the top and base of the Cover Sands. In the more extensive exposures, the Cover Sands are a relatively uniform succession of thin beds of yellowish orange (Munsell No. 10YR6/6) to yellowish brown (10YR4/2) loose sands, with occasional thinner laminae of darker grey (5Y3/2) sand with evident organic debris. Figures 8.1 and 8.2 record typical sections in pits at Messingham (SE 911037 and SE 925037 respectively) and similar sections occur at the second sampling site, in the Ironstone quarries near Flixborough (SE 900154). The thin beds, usually less than 50mm thick and often only apparent on a wind-eroded vertical face, are traceable for considerable distances within exposures but cannot be teleconnected between sites as each eventually runs out, often passing into shallow channel deposits or into sands with distinct ripples or thin cross-bedding. Some layers show graded bedding and other features consistent with water action during deposition. In the Messingham pits, for example, coarse false-bedded units, up to 300mm in thickness run out at right angles to the scarp, gradually attenuating and ramifying westwards into channels among thinner bedded parts of the sands. Frequently, sedimentary features are picked out by iron deposition, due to podzolisation, and

Figure 8.1. Section in Cover Sands (Messingham Sands) at Messingham, S. Humberside (SE 911037) 1981.

rootlet casts similarly defined are also common in most sections. Thin silty organic horizons (figure 8.3) also exert some control on the redistribution of iron compounds in the sand. Below such discontinuous laminae the sand is often a grey (5YR5/1) colour. This colour is more common close to the escarpment, where much of the succession is sealed by a group of interbedded, slightly contorted, silty laminae (figure 8.4). The Cover Sands vary in thickness from a sandy component in the topsoil on the crest of the Wolds (Straw, 1963) and the Isle of Axholme, to in excess of 7m at several sites on the scarp and dipslope of the Lincolnshire Limestone (Elford, pers. comm.). Their altitudinal distribution ranges from below sea level in the valley of the Trent to about 150m O.D. against the Wolds. On Crosby Warren (SE 821127), the blanket of sand sweeps up the Lias escarpment, with depositional dips in excess of 5°, attenuating from nearly 3m to less than 1m in thickness. Beyond the crest, the recently stabilised dunes of Risby Warren and the adjacent reclaimed farmland are of Cover Sands.

From organic horizons in the upper part of the sands on Crosby Warren, Holland (1975) obtained three ^{14}C dates: 2285 ± 70 B.P. (UB-860), 2070 ± 50 B.P. (UB-859) and 1640 ± 435 B.P. (UB-862), which show that movement of sand occurred in the Late Iron Age and Roman periods. These recently reworked sands differ from the Late Glacial Cover Sands since they form dunes, which have not been noted in the older

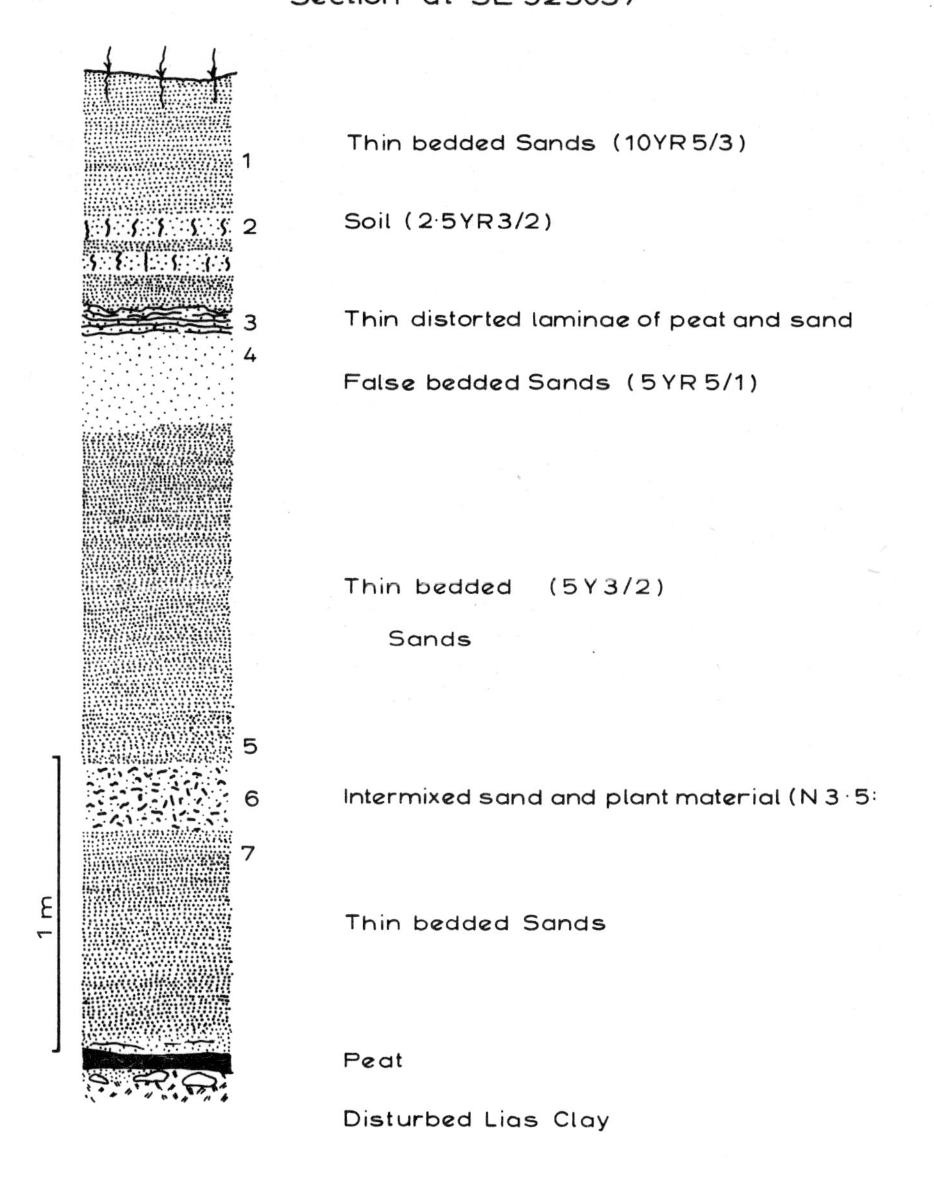

Figure 8.2. Messingham sand quarry, S. Humberside: Section through Messingham Sands at SE 925037, 1973.

Figure 8.3. Messingham sand quarry, S. Humberside, 1981. Section at SE 911037 showing basal peat.

Figure 8.4. Silty organic laminae close to the top of the succession, at Messingham (SE 925037) 1973.

deposits in South Humberside, although they do occur near York (Matthews, 1970).

At the base of the Cover Sands, both in the Messingham pits and in the more extensive exposures of the quarries on Crosby and Flixborough Warrens, a compact layer of peat occurs. This bed, seldom more than 200mm thick, conforms to the gentle dip of the land and extends for a short distance up the scarp slope. At Messingham, the peat can be traced throughout the workings, extending to over $3km^2$. At least $5km^2$ of peat once overlay the quarrying area north of Scunthorpe. Dakyns *et al.*, (1886), Ussher (1890) and Jones and Gaunt (1976) record peat in a similar position elsewhere in the Vale of York and Humberside. The sand pits at Scotter (SE 876036) and Haxey (SE 745005) also contain peat but, as they are worked flooded, the stratigraphic position of material recovered by the grab is in doubt.

A flint end scraper (figure 8.5) and a Bovid astragalus were found in the peat bed at Messingham and a shouldered point together with an unworn Mousterian hand-axe were recovered from a similar stratigraphic horizon on Risby Warren (May, 1976). A sample of the peat from Messingham (SE 915037) gave a ^{14}C date of 10 280 ± 120 B.P. (Birm. 349). A further date was obtained 2m above this compacted peat on a less well developed organic layer, consisting largely of the remains of mosses (*Drepanocladus* spp. Dalby, pers. comm.). This date, 10 550 ± 250 B.P. (Birm. 707), appears older than that from the lower horizon, but the overlapping standard deviations of the dates suggest that this reflects the relatively rapid accumulation of the Cover Sands. The possibility of hard water error, giving too old a date for the upper horizon, should also be considered (Shotton, 1972 and pers. comm.). The dates, however, are closely comparable with others obtained on similar deposits elsewhere in the Vale of York (Gaunt, 1981). Discontinuous layers of dark, slightly organic sand, often with moss remains, occur within much of the Cover Sands succession, but the horizon dated at Messingham is the only extensive one. It can be traced throughout the Messingham pits and there is a similar layer in the same relative position in exposures at Flixborough. On Risby Warren, Dudley (1949) recorded a similar division of the sands, although lacking the basal peat bed, but, in

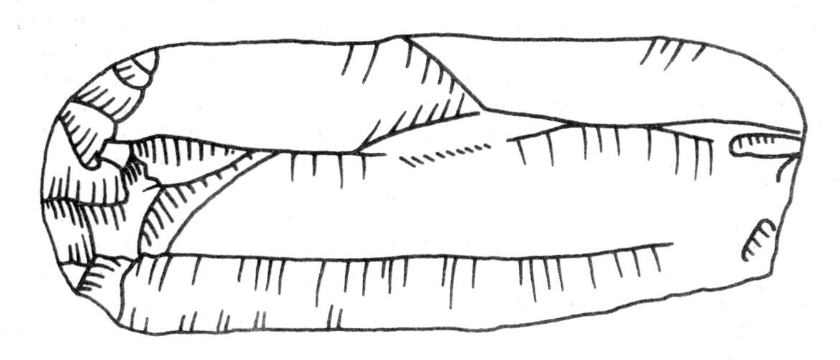

Figure 8.5. Flint and scraper from basal peat at Messingham, S. Humberside.

view of Holland's (1975) Iron Age and Roman dates from Crosby Warren, it would be unwise to correlate over the escarpment. In part of the Messingham pit, close to the base of the escarpment, a succession of silty laminae, rarely in excess of 1mm in thickness, with occasional plant fragments, occurs close to the top of the Cover sands (figure 8.4). These beds have been thrown into a series of irregular ripples either by frost action or by slip down the scarp, and form a unit totalling 200mm in thickness, attenuating westwards until it disappears at about 250m from the scarp. The unit is overlain by about 100mm of slightly iron stained (10YR5/3) sands and a brown (2.5YR3/2) soil, in part divided by blown sand and capped by 500mm of more recent dune deposits. The soils, where sealed by later dunes, are less strongly podzolized than those of the current land surface.

To the west, on Hatfield Chase, peat beds occur within sands over 'Lake Humber laminated clays' at Epworth Turbary (SE 754037) and extensively at Sandtoft (SE 642056) but comparison with sections seen during the construction of the M18 motorway at West Moor, Armthorpe (SE 642056) and the evidence of fossil insect faunas from the peat (Buckland, Coope and Gaunt, in prep.) suggest that these beds belong to the Windermere Interstadial. At Armthorpe, the peat is disturbed by frost action and overlain by a solifluction deposit and it may be the lateral equivalent of the nearby buried soil dated to 11 100 ± 200 B.P. (N-810) (Gaunt, Jarvis and Matthews, 1971).

The Cover Sands, including the underlying peat bed overlie a wide variety of substrates. Frequently they rest directly on the Triassic or Jurassic rock, although this has often been subject to considerable disturbance. At Haverholme, near Appleby (SE 52125), a borehole for the British Steel Corporation proved a depth of 2m of sand and broken limestone beneath 7m of Cover Sands (Elford, pers. comm.). In the Messingham pit, the top of the Lower Lias, largely below water level, is disturbed and a remanié deposit of pebbles and cobbles, including some flint, frost shattered *in situ*, occurs extensively beneath the peat. Several of the sandstone erratics in this discontinuous horizon show evidence of faceting by wind action, a feature noted elsewhere in the Vale of York (Edwards, 1936; Gaunt, 1976). At one point, the peat grades laterally and vertically into a grey (N4) silt with abundant freshwater mollusca and ostracoda, suggesting that a small pool existed before sand deposition began. A sample from this silt produced willow (*Salix* sp.) charcoal. At Flixborough, the Cover Sands overlie extensive pockets of hard panned, poorly sorted gravel, with occasional large erratics of Jurassic calcareous nodules, flint and Carboniferous and Jurassic sandstones. Where structures are discernible, the gravel and adjoining bedrock are seen to be heavily disturbed, with wedges of gravel penetrating down into the Lias clays for up to 2m. It was presumably from these gravels that the teeth and tusk of mammoth, recorded by Dudley (1949) from Flixborough Warren and Thealby, came.

On Hatfield Chase and in the lower part of the Vale of York, the Cover Sands overlie rocks ranging from the Keuper to the final deposits of Lake Humber. North of the York Moraine, Matthews (1970) found sand beneath a peat dated to 10 700 ± 190 B.P. (N-488). These filled frost wedge pseudomorphs in the underlying till. Straw (1963) also found sand beneath the peat on Crosby Warren while at Caistor, it filled fissures, probably frost wedge casts penetrating the rockhead. It is uncertain, however, whether these relate to events earlier in the Loch Lomond Stadial or to previous phases of frost action in the Devensian. Indisputable frost structures have yet to be found in the Cover Sands or the basal peat.

As Straw (1963) recognised, the term 'Cover Sands' included both Late Devensian and Holocene deposits, and some subdivision has inevitably become necessary as a result of further research. Although there remain problems in mapping the Late Glacial succession as distinct units, the formation name 'Messingham Sands' is proposed, to include the deposits from the base of the peat to the surface of the overlying palaeosol, where present. The later Flandrian dunes are specifically excluded. The succession is typified by figure 8.2 (SE 925037) from the Messingham sand pit.

PETROGRAPHY

Although no detailed study of the petrography of the Messingham Sands has been made, a number of points germane to the palaeoenvironmental interpretation have been examined. Microscopic examination of the loose sands shows that they consist of rounded to subangular grains (figure 8.6), principally of quartz, most of which lie in the size range, <0.5mm>0.125mm. Occasional lines of much coarser material occur, however, ranging up to 2.0mm in diameter. The particle size analyses (figure 8.7) confirm the initially aeolian nature of the deposits. Particle size analysis of samples from Messingham, from a section close to the base of the escarpment (figure 8.2) shows that the sands are well sorted. With the exception of the finer grained sample 3, all have over 90% of the sand in the range 0.125-0.5mm. Sample 3 is less well sorted, with all less than 0.5mm and over 90% between the limits 0.5-0.063mm. The cumulative curves (figure 8.7) contrast with the samples from the Old Winteringham (SE 949214) fluvio-glacial deposits but show the sands to be slightly less well sorted than those analysed by Matthews (1970). This results partly from mixing of sand from several laminae in a more variable succession than Matthews studied, but the stratigraphic evidence for partial reworking by water suggests that this may have modified the particle size distribution of the sands. The mid-Devensian Chelford Sands of Cheshire (Harrison, 1968) resemble the Messingham Sands in degree of sorting and a partial reworking of aeolian sands has been suggested in their case (Evans *et al.*, 1968). It should be noted, however, that the samples from the more recent dune (samples 1 and 2) are also less well sorted than Matthews' (1970) material, although there can be no doubt about the wholly aeolian nature of this structure. Similar results were obtained by Wilson, Bateman and Catt (1981) from a more detailed study of the Shirdley

Figure 8.6. Photomicrograph of sand grain from the Messingham Sands at SE 911035.

Hill Sand in Lancashire, although the sedimentary evidence for reworking by water was not evident.

Although the small number of samples examined may not be representative of the entire formation, the variations in grain size between samples seem to have some significance. The curves (figure 8.7) from the recent dune and soil (samples 1 and 2) are similar to that for sample 5, immediately over the horizon which, elsewhere in the quarry, provided faunal sample 2. The finer grained sample 3 represents the highest unit within the undisturbed formation to provide faunal evidence, the fine sands being interbedded with discontinuous organic laminae. This group probably was laid down at the end of the principal period of aeolian activity. Certainly the grain size distribution suggest more influence by water deposition, and the deposit may have been almost entirely resorted. The sand which seals this horizon is similar to samples 1 and 2 and may have been deposited at the base of the scarp in the early Flandrian, before the establishment of vegetation cover. The remaining samples,

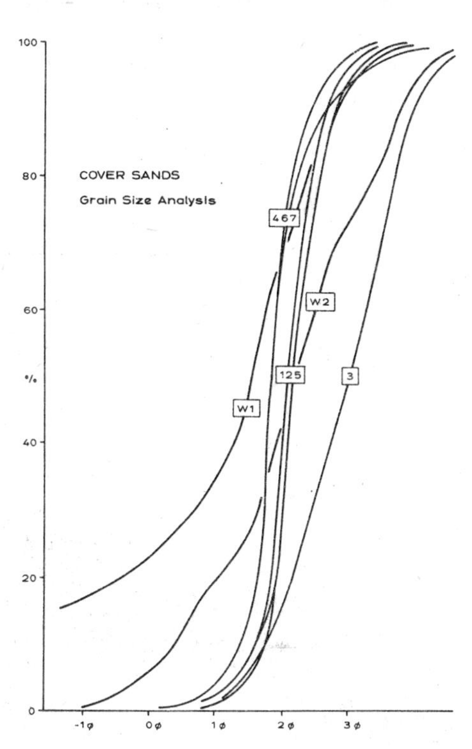

Figure 8.7. Particle size analyses from Messingham Sands etc.

4 (from the cross-bedded unit beneath 3), 6 and 7 (intermixed with and below the organic layer within the body of the sands) have particle size distributions suggesting a more turbulent regime and it is within these that the coarser lines of sand possessing grains over 2mm in diameter, occur.

Despite the granulometric and topographic evidence for an aeolian origin, individual grains show variable rounding under the microscope (figure 8.6). The predominant mineral, quartz, ranges from angular grains with slightly rounded edges to classic millet-seed grains, the latter still retaining their brick-red coating of iron oxides, which implies an origin in the Sherwood Sandstone Group, on the western fringe of Hatfield Chase. The majority of grains, however, are less distinctive, most being sub-rounded and lacking evidence of origin. Occasional sub-rounded fragments of fine-grained igneous and metamorphic rocks occur, as well as subangular fragments of flint of more local derivation. The Chalk and Permo-Triassic outcrops lie on either side of the principal outcrop of the Messingham Sands, however, so that it can be postulated either that the depositional winds blew from both east and west or that the flint and/or Sherwood sand were obtained via an intermediate erosional cycle. The distribution of cover sands (figure 8.8) implies deposition by winds blowing across the Lower Trent Basin and the Vale of York, where the exposed sands and gravels of Late Devensian Lake Humber and the Triassic outcrops would provide the rock types indicated. The surface of the Older Drift around Wroot and Finningley, south-east of Doncaster, has the appearance of a remanié. The finer fraction must have contributed to the formation of the Messingham Sands, and a wind direction slightly south of west would increase

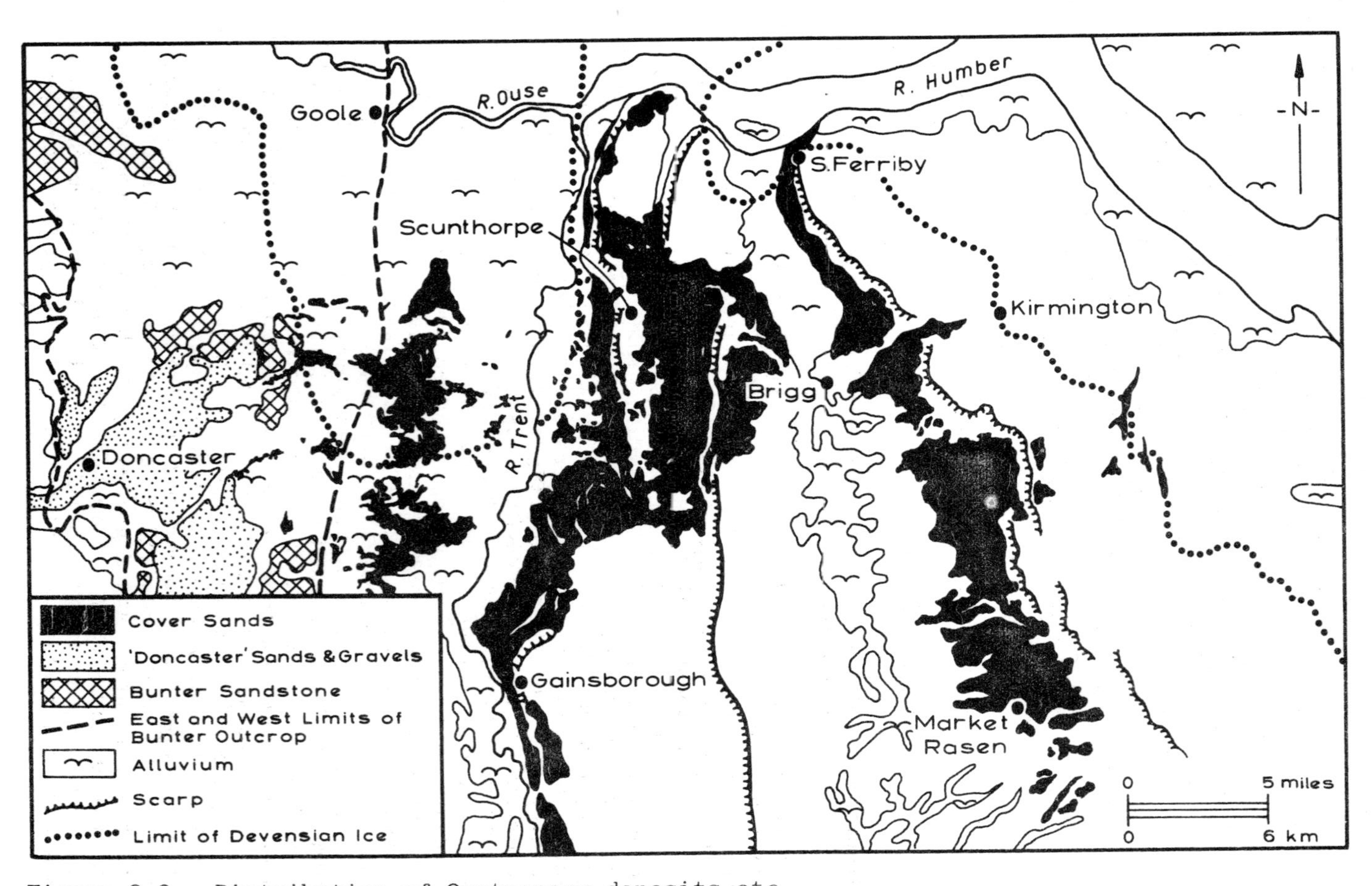

Figure 8.8. Distribution of Quaternary deposits etc.

the possibility of inclusion of millet-seed grains, since the Sherwood Sandstone Group outcrop widens considerably into Nottinghamshire. Dunes and spreads of sand occur in the Vale of York in the same stratigraphic position as the Messingham Sands and these have similar implications for wind direction, although the evidence from dunes is debateable. Such a model need not contradict the impressive evidence assembled by Sissons (1980) for a relatively dry climate during the Loch Lomond Stadial, provided it is stressed that the Messingham Sands episode represents the rapidly changing conditions at the very end of the Stadial, perhaps the 'transition' which Lowe and Gray (1980) place between 10 500 and 10 000 B.P. A similar division of the period has been suggested on limnological grounds by Pennington (1975) from northern Britain but it should be noted that the insect evidence implies cold conditions at least until the cessation of deposition of the Messingham Sands in Lincolnshire. West of the Pennines. the stratigraphically similar Shirdley Hill Sand of south-west Lancashire provides no evidence for a dominant wind direction (Wilson, Bateman and Catt, 1981). Westerly winds have been suggested for Younger Dryas Cover Sands in the Netherlands (Rutten, 1954; Maarleveld, 1960) but de Jong (1967) suggested north-westerly winds during both Older and Younger Dryas times.

The efficiency of the wind in rounding sand is uncertain (c.f. Pettijohn, 1957) and the varying degrees of roundness exhibited by the Messingham Sands reflect various origins rather than their Late Glacial history. The small proportion of subangular flint fragments was probably derived subaerially from outwash in the Vale of York but some could have been incorporated at Messingham by washing down the slope from tills and sand and gravel pockets along the escarpment.

Heavy mineral analysis has not been employed to examine the possible sources of the sand but Loughlin (pers. comm. and 1977), in his search for the manufactories of Dales Ware, a common Roman pottery type, noted frequent subhedral crystals of apatite in sand from Messingham. Smithson (1931) found that this mineral is common in the Trias of Yorkshire, particularly in the Keuper, and this is extensively exposed on the Isle of Axholme, 12km west of Messingham. Much more work is needed on the petrography and origin of the Messingham Sands.

SAMPLING FOR INSECT REMAINS

The casual find of a flint end scraper in the peat on the floor of the Messingham quarry provided the impetus to examine the Messingham Sands closely. The flint industries of the Late Glacial, particularly the Creswellian, are poorly defined, with few securely dated sites in Britain, and their relation to the succession of rapidly changing environments has been little explored (Campbell, 1977). The connection between scraper and peat bed and the later discovery of charcoal in one sample from the basal horizon gave an opportunity to attempt an environmental model for the implement's user at one moment in time (10 280 ± 120 B.P.), at the very end of the Palaeolithic.

Samples were initially taken in the Messingham pit close to the find-spot of the scraper (SE 915037). A Bovid astragalus, misidentified as *Megaloceras giganteus* in Buckland (1976), was also found in the peat near this locality but there was no direct connection with the artifact and, as one of the most durable bones in the body, it could be residual from earlier in the Late Glacial. 5kg samples were taken from the hard, compressed fibrous peat on the quarry floor in two arbitrary 50mm divisions and samples were also recovered from a number of slightly organic lenses and laminae within the Messingham Sands. Unfortunately, the latter failed to provide any insect remains but, later, as the working face progressed eastwards, the principal organic horizon within the Sands was better developed and the moss sample which gave the ^{14}C date of 10 550 ± 250 B.P. (from SE 918039) also yielded a limited insect fauna. Close to this locality, the basal peat passed laterally into a more silty deposit, which contained a limited insect fauna, many ostracods and, probably of archaeological significance, a few fragments of willow charcoal. Later, higher laminae with insect remains were found close to the base of the escarpment (SE 925037). North of Scunthorpe, the Ironstone mines provide abundant sampling locations and it was possible to select a section on Flixborough Warren (SE 900154) where the basal peat and the intraformational organic horizon were both well developed, although their faunas were rather limited.

All the samples from within the Messingham Sands consisted of loose aggregations of moss, identified as predominantly *Drepanocladus* sp. (Dalby, pers. comm.). These were easily washed out on a 300 μm sieve and sorted in their entirety for insect remains. The compact, felted peat from beneath the sand unit, however, presented more of a problem. This contained much moss and also fragments of the leaves of grasses and sedges, the nutlets of the latter being frequent. There was little woody tissue although a few small twigs and a leaf of *Salix*, probably *herbacea*, were noted whilst splitting the peat. Samples were broken down in frequent changes of hot sodium carbonate solution and attempts were made to split the peat along its irregular bedding planes. After about ten days of this treatment, it was possible to wash each sample through a 300 μm sieve. The material retained on the sieve was then subjected to paraffin (kerosene) flotation (Coope and Osborne, 1968). Although a good separation of an insect-rich floatant was obtained, large amounts of indeterminate plant debris also floated and the insect fragments had to be laboriously culled from this. The two samples of the basal peat at Messingham yielded similar faunas and are therefore combined in the faunal list (figure 8.9) given at the end of this paper.

Preservation of insect remains was variable and slides prepared for pollen analysis from the basal peat at Messingham produced only crumbled grains of *Artemisia* and *Thalictrum*. Koster (pers. comm.) has obtained an asemblage similar to that of Jones and Gaunt (1976) from an organic horizon within the sands at Messingham (SE 911039) and Taylor and Baxter (pers. comm.) prepared successful slides from a locality a kilometre to the north (SE 912047). More detailed palynological work on the organic horizons of the Messingham Sands would be rewarding.

NOTES ON PARTICULAR SPECIES

Late Glacial insect faunas in Britain have been extensively studied by Coope and Brophy (1972), Coope and Joachim (1980), Osborne (1972), Ashworth (1972; 1973)* and others and several species have been discussed in detail (Coope, 1966; Angus, 1973). Therefore, only those beetles which are additions to the Late Glacial list and those of particular interest in the Cover Sands faunas are considered here.

Bembidion velox

A single elytron of the *Bembidion* subgenus *Chrysobracteon* was recovered from sample 3, from Messingham. Both size and microsculpture separate this individual from the two British members of the group and similarly eliminate other Continental species, except *B. lapponicum* Zett. and *B. velox* (L.). The general habitus suggests the former species but the width of the third interstice, compared with the second, is regarded by Lindroth (1940) as diagnostic and this places the specimen firmly within *B. velox*. This is a stenotopic species, found on damp, pure, sterile sand, often by lakes and rivers (Lindroth, 1945). It is widely distributed in North and Central Europe, extending eastwards across Siberia, but it is rare and there are no more recent records from southern Germany and Austria than 1920 (Freude, 1976).

Trechus rivularis

Superficially, the cryptic fenland ground beetle, *T. rivularis* would seem an improbable member of Late Glacial assemblages. Lindroth (1945; 1974) suggests that it is strongly stenotopic, restricted to 'dark forest swamps with *Sphagnum* among damp sedge litter.' In Britain, it is known only from the Cambridge and Norfolk fens and Askham Bog, near York, an isolated relict fen. Neither the latter locality nor its most prolific Fenland site, Wicken Fen, can really be described as 'forest swamp' and the damp sedge litter, with minimum disturbance and reasonable shade not necessarily from trees, are perhaps the most important factors, the present rarity being related to anthropogenic factors (c.f. Buckland, 1979). As well as a Hoxnian record (Shotton and Osborne, 1965), there are several Devensian records from phases with little or no evidence of tree cover - Four Ashes (Morgan, 1973) and Rodbaston, Staffordshire (Ashworth, 1973), the Tame Valley, Warwickshire (Coope and Sands, 1966) and Colney Heath, Hertfordshire (Pearson, 1961). In the early Flandrian, the species is recorded from Rodbaston (*op. cit.*), Lea Marston, Warwickshire (Osborne, 1974), and Red Moss, Lancashire (Ashworth, 1972). Despite Pearson's

* For full bibliography, see Buckland (1981).

(1961) optimistic distribution map (compare Lindroth, 1949; Hansen *et al.*, 1960), *T. rivularis* is not recorded from the tundra and Lindroth (1945) does not note it beyond his 'regio coniferina'. It is possible that the species bears ecological similarity to the weevils *Barynotus squamosus* and *Otiorhynchus nodosus* discussed by Coope (1965) and is able to avoid direct sunlight by being nocturnal in low latitutde near-tundra situations, an environment with no direct modern analogue. In the Far North, it would be unable to achieve this because of increased summer day-length. Its present distribution, however, suggests that it is unable to take advantage of the more extensive cloud cover of oceanic areas in order to be active during the day and other factors, perhaps wetter and warmer winters, must exclude it from the West.

Bledius spp.

The wet, sandy substrates provided by the Messingham Sands formed ideal habitats for species of *Bledius* and several are represented, although taxonomic and other problems preclude firm identifications. The rare north European *B. vilis* Makl. was tentatively identified on thoraces and elytra, using collections in the Manchester Museum. Comparison with fossil specimens from mid-Devensian deposits at Upton Warren, Worcestershire, however, suggests that these specimens may belong to *B. littoralis* Heer, identified by Tottenham (in Coope *et al.*, 1961) from the latter site. Both species are northern in their distribution and identification of the Messingham Sands material must await the recovery of more diagnostic fossil remains.

Philonthus punctus/binotatus

Although Reitter (1909) and Horion (1951) synonomise these species, they are separated by both Palm (1952) and Coiffait (1967). Despite clear distinctions in the aedeagi and thoraces, there would appear to be some overlap in the shape and puncturation of the heads, at least in the males, and the species cannot be separated on the Messingham material. The synonomy makes any attempt to draw conclusions from the apparent distributions difficult, but *P. binotatus* is recorded from Skåne, at the southern tip of Sweden, and the Baltic coast of Holstein; Lohse (1964) suggests that it may be a halophile. *P. punctus* is apparently more widespread, occurring, if somewhat rarely, throughout Central Europe and eastwards to the Caucasus and Turkestan (Horion, 1951). Further north, the beetle tends to be rather southern and is restricted to southern England (Joy, 1932), absent from Norway and restricted to south of Lat. 61°N. in Sweden and Finland (Hansen *et al.*, 1960). Joy (1932) regards the insect as a coastal species but continental sources range from Reitter's (1909) common in woodland, by flowing sap on trees, to Lohse's (1964) uncommon, on banks of large lakes and rivers. Insects which are often regarded as halophiles occur in other Late Glacial assemblages (e.g. *Ochthebius marinus* from Glannllynau (Coope and Brophy, 1972) and they probably relate to the undeveloped nature of the soils prior to post-Glacial leaching. The present distribution of these species suggests that they are relatively thermophilous but *P. punctus* is also recorded from a very similar cold fauna from the Tame Valley at Minworth of Devensian age (Coope and Sands, 1966) and other factors, probably related to the degree of continentality, are clearly involved.

Aphodius obscurus

This dung beetle was previously recorded from deposits of 11 790 ± 140 B.P. - 12 135 ± 200 B.P. (N.P.L.81, Birm 158) age at Church Stretton, Shropshire (Osborne, 1972). Its distribution at the present day is both southern and continental, occurring in sheep, goat, chamois and cattle dung in the Central European mountains, the Pyrenees, Alps, northern Appennines, Anatolia and the Caucasus (Balthazar, 1964). Such a pattern suggests a preference for a high alpine environment but Machatschke (1969) also notes the species from the lower areas of the Black Forest and Thuringerwald. This could imply that *A. obscurus* is relatively thermophilous and Osborne's (1972) record is not associated with a particularly cold assemblage. Although it is generally thought that day-length is not a significant factor is insect distribution (Downes, 1965; Morgan, 1973), dung beetles, with their propensity for nocturnal flight activity (Landin, 1961), may relate to both macro- and microclimatic parameters associated with this. It is impossible to ascertain, without extensive experimentation, at what stage in an insect's development a climatic factor becomes limiting, but a cold stenotherm could occur in the areas occupied by *A. obscurus* if it required a combination of high daytime insolation for larval development and cold nights for imaginal dispersal, a regime engendered by katabatic airflow from surrounding mountains.

*Aphodius montivagus**

Heads, thoraces and an elytron were recovered from Flixborough and Messingham of a distinctive species of *Aphodius*. Using Machatschke's (1969) key, the small relative size of the head, absence of a posterior margin, and the puncturation of the thorax placed these individuals in the subgenus *Aegolius*. Size eliminated several species but left a group of four, in which *A. montivagus* could be differentiated on elytral puncturation and microsculpture. Despite problems with the taxonomy of the subgenus (c.f. Nicholas, 1971), the identity was confirmed by specimens from continental collections. Unlike the majority of *Aphodius*, *Aegolius* spp. feed on decaying plant material and all are predominantly high alpine in distribution (Machatschke, 1969). *A. montivagus* is widespread in the Calcareous Alps, Styria and Carinthia, Austria; a French record (subsp. *cenisius* Dan.) (Paulian, 1959) is omitted by Balthazar (1964). The comments on diurnal rhythm for *A. obscurus* may be equally relevant to this species, although it appears to be a more specifically high altitude insect and the problem of eco-

* a thorax of this species was tentatively identified as *A. (Neaegolius)* prob. *piceus* (Gyll.) in Buckland (1976).

logical interpretation are similar to those of the Tibetan *A. holdereri* Reit. from mid-Devensian sites (Coope, 1973).

Notaris bimaculatus

One elytron and a thorax of a weevil from Messingham (sample 1) could not be matched in the British Museum's collection of Coleoptera. Although its Notarine affinities were fairly evident, it was much more strongly regulose and shining in its interstitial ornamentation, smaller than any species of *Notaris* and larger than most *Thryogenes* spp. M. Girling (pers. comm.) found six undescribed specimens in the Collection, however, standing under the generic name *Lixellus*, which compared very closely with the fossil. The modern individuals, from Nova Scotia and Hudson's Bay, Canada, were subsequently determined by R.T. Thompson (pers. comm.) as lying within the range of variation of the Holarctic species, *N. bimaculatus*, also recorded, in its more usual form, from this and other Messingham Sands samples.

THE PALAEOENVIRONMENT

The Basal Peat

The close similarity between the faunas from the peat at Messingham and Flixborough implies a similar environment at both localities and over much of the Lias dipslope during the Loch Lomond Stadial. At Messingham, an extensive water beetle fauna occurs and several species suggest a pond with open water. The Dytiscids include large, free-swimming species of *Ilybius* and *Agabus*, with *Rhantus exsoletus* and *Colymbetes* sp. *Agabus arcticus* prefers small, shallow pools of standing water, frequently with Sphagnum (Lindroth, 1935; Balfour-Browne, 1950). The Hydrophilids occupy thickly vegetated margins to ponds, although *Chaetarthria seminulum* is found in wet mud, away from vegetation, at the edge of pools (Balfour Browne, 1958). *Helophorus sibiricus* shows an interesting variation in habitat across its range. In Scandinavia, it is found by the sides of mountain streams and rivers but Siberian records are from grassy pools, particularly those by melting snow (Angus, 1973), and it is the latter habitat which accords best with the other faunal evidence from Messingham. Much of the fauna is associated with decaying plant debris, including the large number of *Cercyon* spp. and several species of Staphylinid. Mosses were apparent in both collection and preparation of the samples and with these are associated the Byrrhids, *Coelostoma orbiculare* and Scirtids, as well as many of the Staphylinids, including the most frequent taxa in both major samples *Olophrum fuscum* and *Arpedium brachypterum*. The Carabids, *Diachila arctica* and *Elaphrus lapponicus*, are also recorded from wet mosses and both are strongly hygrophilous (Lindroth, 1961). Among the other ground beetles, the association of *Trechus rivularis*, *Bembidion doris*, *Pterostichus diligens* and *Agonum fuliginosum* is typically that of a shaded *Carex* marsh (Lindroth, 1945), although *B. schuppeli*, also relatively common in the principal

Messingham sample, is more riparian, preferring sand mixed with plant debris and sparser vegetation (Lindroth, 1974). Patches of a more sandy substrate do occur beneath the peat bed and occasional intercalations of sand were noted towards the top in a number of localities and the specimens may be associated with such areas.

The evidence from phytophagous insects for the plants growing in this marshy environment is relatively limited. This faunal element is dominated by taxa associated with the Cyperaceae, whose seeds were abundant in both main samples. Although Joy (1932) gives the bur-reed, *Sparganium erectum*, a species with a fairly southern modern distribution (Fitter, 1978), as the food plant of *Notaris aethiops*, this weevil, the most common in the Messingham sample, is probably more polyphagous and could find suitable hosts among the sedges. Pearson (1961) suggests *Carex* spp. and Osborne (1973) has *Scirpus* spp. Its congener, *N. bimaculatus*, has been recorded feeding upon *Typha latifolia* and *Phalaris arundinacea* (Hoffman, 1958). The other Notarine weevils in the sample, *Thryogenes* spp., feed upon various species of *Carex* and *Scirpus* (op. cit.). *Limnobaris pilistriata* occurs on various sedges and *Juncus effusus* (Hoffman, 1954) and the reed beetle, *Plateumaris sericea* is also largely associated with the sedges (Stainforth, 1944). Evidence for floating vegetation on ponds at Messingham is provided by *Phytobius canaliculatus*, which has been recorded from the pondweed *Potamogeton natans* and, where their hosts are known, species of *Bagous* are found on floating aquatic plants *Gryphus equiseti* appears on the marsh horsetail, *Equisetum palustre*. The species of *Otiorhynchus* are polyphagous, although *O. rugifrons* seems to prefer *Thymus serpyllum* (agg.) in Iceland (Larsson and Gígja, 1959), and *Sitona suturalis* has been recorded from several species of *Ononis, Vicia* and *Lathyrus* (Hoffman, 1954). The small Chrysomelid *Phaedon* sp. (- *P. tumidulus* can be excluded -) appears on a variety of plants in damp situations (Mohr, 1966), usually on Umbelliferae (Joy, 1932).

Despite the abundance of birch pollen in Zone III at Aby Grange, south-east Lincolshire, (Suggate and West, 1959), and at Tadcaster, Yorkshire (Bartley, 1962), and macroscopic records which extend into north-east Scotland (Godwin, 1975), there is neither macroplant nor insect evidence for the presence of this tree in the basal peat in north Lincolnshire. Some wood is implied by *Anaspis* sp., as the larvae of most species develop in rotten wood, but this habitat could have been provided by the willows, for which there is good evidence. As well as the charcoal from the underlying sample and the leaf of *Salix*, probably *S. herbacea*, found in a Messingham sample, two individuals of the small weevil, *Apion minimum*, were identified. This has been recorded from several types of willow, developing in the leaf galls of various Tenthredinid sawflies (Hoffman, 1958). Some of the Staphylinids, particularly *Arpedium brachypterum* and *Boreaphilus henningianus*, may also occur in willow leaf litter. From Flixborough, an example of *Phyllodecta* can be added to the list of willow feeders, although some species are also found on poplars (Mohr, 1966).

Several of the species of Hydrophilid and Staphylinid associated with decaying plant debris may also occur in dung and this is also indicated by *Aphodius* spp. *A. merdarius* is predominantly an open ground insect, recorded from horse and cattle droppings (Landin, 1961), and *A. obscurus* has been found in sheep, goat and chamois dung (Balthazar, 1964). Of these, only the horse is known to have been in Britain during Zone III (Grigson, 1978). Dung beetles, however, are not tied to particular herbivores but to the character, situation and microclimate of the droppings. Both species could have occurred in reindeer dung, since this animal is the most frequent large vertebrate recovered from Zone III deposits, or in that of 'tundra bison', *Bison bonasus arbusto-tundrarum* Deg., recorded at this period in Denmark but not so far in Britain (*op. cit.*). It is tempting to relate the one bone found at Messingham to this species but, unfortunately, the bone, a Bovid astragalus, is insufficiently diagnostic (Rackham, pers. comm.). The other species of *Aphodius* identified, *A. montivagus*, is a plant debris feeder rather than coprophagous.

The image of the pre-Cover Sands landscape on the Lias dipslope is therefore relatively complete. An extensive *Carex*-dominated marsh, with some scrub willow and occasional semi-permanent pools, probably fed by meltwater from local snow patches, forms the backdrop against which the users of the flint scraper from Messingham and shouldered points from Risby must stand. Their prey, probably horse or reindeer, picked their way across the marsh, drinking from the scattered pools, by which man, the hunter, occasionally crouched with a few dry twigs of willow for a fire to warm himself against the cold winds of the latter part of the Younger Dryas. Caution must be exercised, however, in extending this picture. Charcoal is frequent in the 'Usselo layer' of the Netherlands, of late Allerød or early Younger Dryas age (Maarleveld, 1960) and the Messingham loose association of scraper and charcoal could provide support for an anthropogenic origin for this more extensive phenomenon, as charcoal from Allerød soils in southern England has (Evans, 1975). Natural fires are a potent factor in the maintenance of sub-climax communities in conifer woodland in North America (Wright & Heinselman, 1973) and lightning strikes into moribund pinewoods, a result of climatic deterioration, seem at least an equal possibility. Less convincingly, Paddaya (1972), after Waterbolk (1954), has blamed ash from volcanic eruptions in the Eifel for the burning of pine forests in the Late Glacial of the Low Countries.

Away from the extensive marsh, stands of birch and scrub of willow and dwarf birch must have existed in favourable localities but the insect evidence from the north Lincolnshire sites implies a landscape of unconfined arctic mires, similar to the muskeg of the Canadian and Alaskan tundra (Bellamy, 1972). Despite the evidence for solifluction and cryoturbation during Zone III elsewhere, convincing frost structures have not been found in either the peat or overlying Cover Sands.

The Cover Sands

Although occasional thin laminae of sand occur towards the top of the basal peat in several section (figure 8.3), the transition to a phase of widespread sand deposition is relatively abrupt and it is clear that the vegetation cover was broken over a very wide area in a short space of time by extensive aeolian action. Organic horizons are less frequent in the first 2m of sands, the rare stringers of slightly organic material consisting, where recognisable, of moss fragments. It was initially thought that the succession of thin, sub-horizontal sand beds and laminae represented an annually deposited sequence but, as the quarry face at Messingham advanced towards the base of the escarpment and the Crosby and Flixborough sections were examined, it became evident that water had redeposited much of the sand and the sequence was less chronologically simple. The character of the deposits, however, and the overlapping, inverted ^{14}C dates imply rapid accumulation.

When faunas reappear in the Messingham Sands, at a horizon traceable in most sections, roughly 2m above the base of the Messingham pits, the phytophagous insects are absent and the beetles are detrital feeders and predators associated with a sandy substrate. Species of sand-burrowing Staphylinid, *Bledius* spp., appear and the only ground beetle is the bare ground stenotope, *Bembidion velox*. It is probably the edaphic transformation and paucity of litter which reduce the numbers of *Arpedium brachypterum*, although the other Omaliine which tends to be co-dominant on cold sites, *Olophrum fuscum*, maintains its relative numerical superiority and other moss-living species, *O. boreale* and *Acidota cruentata* remain. At both Flixbrough and Messingham, some open water is indicated by the larger Dytiscid water beetles, again including *Agabus arcticus*, but the Hydrophilids are seriously reduced and only represented by small species. *Helophorus sibiricus* is supplemented by *H. glacialis*, also a cold stenotherm, and in northern Europe confined to the margins of snow patches on bare ground always near freezing point (Angus, 1973). That large herbivores might still be at least intermittently present is suggested by two specimens of *Aphodius*.

After this more quiescent interlude, perhaps a single warmer summer, aeolian deposition began again and a further one to two metres of sand was laid down against the west facing scarps. Away from the escarpment little organic material has survived but, at the eastern limit of the Messingham pit, a coarse cross-bedded unit of sand is capped by several thin beds of finer sand with traces of organic debris (figure 8.4). These yielded insect faunas, sampled in two arbitrary divisions of 50mm. Apart from the decline in the number of individuals, these remain substantially the same as in the lower horizon in the Sands. The species of *Bledius* are joined by their predator, *Dyschirius politus*, a ground beetle of sparsely vegetated fine sand habitats (Lindroth, 1974), and a phytophagous element reappears, with moss feeders, *Simplocaria* sp., and the weevils, *Notaris aethiops* and *Otiorhynchus nodosus*. The dung beetles may again imply herbivores in the region. No higher organic horizons were found in the Messingham Sands and it is unfortunate that the progression through into the Holocene (Flandrian) could not be followed. Dudley (1949) noted a peat, nearly 2m thick, containing a temperate vertebrate fauna, within sands near Santon (c. SE 930115) but this has been destroyed in ironstone mining.

THE CLIMATE

An attempt must now be made to provide climatic parameters for the environmental model constructed from the Messingham Sands data. The geographical distribution and limited petrographic evidence show deposition by westerly winds blowing across the Vale of York and Lower Trent Valley. But similar conditions need not have prevailed during the period in which the basal peat accumulated so that aeolian sand and peat have to be treated separately. The whole of the small fauna from the peat at Flixborough may be found today in the adjacent provinces of Torne Lappmark and Lapponia Kemensis on the northern Swedish-Finnish border (Hansen *et al.*, 1960), although not necessarily within the same altitudinal range. This region lies within the Arctic Circle but south of the zone of extensive permafrost (Péwé, 1969). In this area lies the northern limit of *Limnobaris pilistriata*, a limit which seems to relate to climate rather than the distribution of its hosts, the Cyperaceae. This northward range overlaps with the southern limit of *Diachila arctica*, which is known from the tundra down into the northern coniferous belt (Coope, 1966). From their present distribution, both *D. arctica* and *Helophorus sibiricus* suggest continental conditions, with considerable contrast between summer and winter temperatures. On its Scandinavian range, *Trechus rivularis*, a more southerly element in the apparently contemporary Messingham fauna, also implies a continental regime, although this species has relict populations in eastern England. Increased continentality of climate would be expected from the low sea level (Jelgersma, 1966) and anticyclonic circulation over the Scandinavian ice cap and this conclusion finds considerable support from the distribution of periglacial features (Williams, 1975). A tendency for the Scandinavian anticyclone to dominate the atmospheric circulatory pattern during the summer months would have kept temperatures low. Coope (1975), from the insect faunas of several Late Glacial sites, has suggested average July temperatures of the order of 9°C and both entomological and geomorphological data combine to suggest January averages of the order of -10°C (Coope, 1977). Absolute contemporaneity cannot be assumed but there is little to contradict these estimates in the north Lincolnshire faunas and Thompson's (1968) account of conditions around Hudson Bay provides a better impression of the contemporary environment than any current European situation. Some evidence for frost disturbance of the peat, such as Moore and Bellamy (1973) describe in Arctic Canada, however, might be expected and this has yet to be found within either the peat or overlying sands in Lincolnshire. Its absence could be explained by postulating deep winter snow cover, since thicknesses in excess of 400mm insulate the ground from frost action and permafrost formation (Brown, 1970). Such insulation is not a feature of most existing northern continental environments, where the tundra landscape is exposed to more rigorous conditions, as the wind sweeps the snow away from large areas. It would, however, explain several of the apparently anomalous, more thermophilous elements in the insect assemblages, such as *Trechus rivularis* and *Aphodius obscurus*.

The initiation of a peat growth which extensively mantles the landscape is somewhat problematic, although the compact nature of the deposit has clearly been instrumental in its continued preservation. A threshold in declining thermal regime may have been crossed, allowing production, largely bryophyte, to dominate over decomposition but deep snow and widespread cessation of active permafrost may have also aided the accumulation of organic debris. The failure of winter renewal of the active layer could have caused a gradual wastage of the frozen groundwater regime, particularly on the more porous formations of the Sherwood Sandstone Group and Drift in the Vale of York and Nottinghamshire. The disappearance of an ice-perched surface wet layer would have allowed westerly winds access to the poorly consolidated sands and sandstones, leading to extensive aeolian action. Black (1951) noted that the presence of permafrost, perching the water table and allowing summer plant growth, prevents the Arctic Coastal Plain of Alaska becoming a cold, arid desert with extensive sand blowing. Aeolian deposition is restricted here to the margins of major rivers (Richert and Tedrow, 1967), where the dunes contain organic lenses, as Late Glacial deposits elsewhere in the Vale of York (Matthews, 1970), but these form a poor parallel for the extensive sheet deposition of the Messingham Sands.

The transition from peat growth to aeolian sand deposition in all exposures of the Messingham Sands is fairly abrupt but some indication of a progressive breaking of the poor vegetation cover is provided by occasional lenses of sand towards the top of the peat (figure 8.3). An erosional contact was not noted in any of the extensive exposures but the possibility of some hiatus between peat growth and sand accumulation, caused by a further decline in available energy resources, remains. As at the present day in the Vale of York, one particularly violent storm may be sufficient to initiate sand movement over a large area (Radley and Simms, 1967). Embleton and King (1975) suggest that wind speeds in excess of 30km/hr are required to move sand in the size range 0.25 - 0.5mm. In the Vale of York and Lower Trent Basin, such conditions prevailed, probably seasonally, until an average depth of about 2m had accumulated against the Keuper, Lias, Lincolnshire Limestone and Chalk escarpments. The presence of occasional organic laminae within the sands, some with sufficient faunal evidence to show continuing cold conditions, and the preponderance of thin bedding suggest rhythmic, possibly annual sedimentation, although partial reworking by water prevents any effective count of years.

The evidence suggests a sequence of summers which were slightly warmer and wetter than earlier in the Stadial, with considerable plant growth, particularly of semi-aquatic mosses, on the wet surface of the sand. Each summer was followed by vigorous storms, sweeping in from the west during the autumn, with deep, early winter snowfall and the ensuing spring thaws partially redistributed the sands. Uniform subhorizontal laminae of sand occur in arid environments in Libya (McKee, 1964) but the biological evidence for wet conditions and waterlogging precludes a similar, if cold origin.

The development of a predominantly continental climate, involving deep snow cover in at least part of England, towards the end of the Loch Lomond Stadial, can be seen in terms of the partial regeneration of the Scandinavian ice cap and the growth of its associated anticyclone, a model discussed by Brooks (1926) and elaborated by Zeuner (1959). During the climatic deterioration of the latter part of the Windermere Interstadial, south-westerly winds, collecting moisture over an ice-free Atlantic (Ruddiman and McIntyre, 1981), would have deposited snow over the ice cap, contributing to its growth or stabilisation. As the anticyclone associated with the ice cap expanded, the frontal systems would have been pushed westwards, causing any moisture-laden westerlies to rise, giving increased snowfall in Britain. Under conditions prevailing earlier in the Late Devensian, this resulted in the development of an ice sheet over northern Britain. In the Late Glacial, this process was short-circuited by an increased thermal input to the westerly airstream, resulting in the rapid destruction of a stable Scandinavian anticyclone, an episode to which the phase of cover sand deposition both east and west (Wilson, Bateman and Catt, 1981) of the Pennines may be related. Although Pennington and Lishman (1971) have shown that interpretation is more complex than was first thought, the rise in iodine levels in Late Glacial deposits in Lake District lakes may also have been controlled by this westerly wind component. The frequent intercalation of organic lenses within the Messingham Sands makes it an ideal formation for the detailed study of climatic change, inferred from biota, at the very end of the last glaciation. Further research may well resolve many of the difficulties surrounding the seemingly abrupt transition at around 10 000 B.P.

Only a hint of a transition to Post-glacial conditions is provided by particle size analysis of the highest unit providing faunal information. In this, the limited fauna remains cold but the westerly storms seem to have declined, the mean frontal position perhaps having moved eastwards, and a less well sorted unit of finer material accumulated, largely by water action at the base of the scarp at Messingham. Work, particularly by Osborne (1974; 1980) on insect faunas from the Midlands and by Bishop and Coope (1977) in Scotland suggests a very rapid transition to conditions probably warmer, if more continental than today.

Figure 8.9 List of insect remains from the Messingham Sands

	Sample number					
	1	2	3	4	5	6
Hemiptera						
Saldidae						
Salda littoralis (L.)	-	-	1	-	-	-
Coleoptera						
Carabidae						
Carabus sp.	-	1	-	-	-	-
Notiophilus aquaticus (L.)	-	1	-	-	-	-
**Diachila arctica* Gyll.	-	1	-	-	1	-
Elaphrus cupreus Duft.	-	1	-	-	-	-
E. lapponicus Gyll.	-	1	-	-	-	-
Dyschirius globosus (Hbst.)	-	3	-	-	-	-
D. politus (Dej.)	-	-	-	1	-	-
Dyschirius sp.	1	-	-	-	-	-
**Bembidion velox* (L.)	-	-	1	-	-	-
B. schuppeli Dej.	1	7	-	-	-	-
B. doris (Panz.)	-	7	-	-	-	-
Bembidion spp.	-	10	-	-	-	-
Trechus rivularis (Gyll.)	1	18	-	-	-	-
Pterostichus diligens (Sturm.)	-	1	-	-	-	-
Pterostichus sp.	-	-	-	-	1	-
Agonum fuliginosum (Panz.)	1	3	-	1	1	-
Dytiscidae						
Hygrotus quinquelineatus (Zett.)	-	1	-	-	-	-
Hydroporus erythrocephalus (L.)	-	1	-	-	-	-
H. melanarius Sturm.	-	-	1	-	-	-
H. palustris (L.)	-	2	-	-	-	2
Hydroporus spp.	-	3	2	-	1	-
Agabus arcticus (Payk.)	-	2	1	-	-	-
A. bipustulatus (L.)	-	1	-	-	-	-
Agabus sp.	-	1	2	-	-	-
Ilybius spp.	-	2	-	-	-	-
Rhantus (?) *exsoletus* (Forst.)	-	1	-	-	-	-
Colymbetes sp.	-	2	1	-	-	1
Hydrophilidae						
Hydrochus brevis (Hbst.)	-	3	-	-	-	-
**Helophorus glacialis* Vill.	-	-	3	3	-	-
**H. sibiricus* Mots.	-	2	5	-	3	-
Coelostoma orbiculare (F.)	-	2	-	-	-	-
Sphaeridium (?) *lunatum* (F.)	-	1	-	-	-	-
Cercyon convexiusculus Steph./ *sternalis* Sharp	1	11	-	-	-	-
C. haemorrhoidalis (F.)	-	3	-	-	-	-
Cercyon spp.	4	14	-	-	-	-
Hydrobius fuscipes (L.)	-	4	-	-	-	-
Helochares (?) *lividus* (Forst.)	-	2	-	-	-	-
Enochrus sp.	-	2	-	-	-	-

cont...

Fig 8.9 cont...

Taxon						
Chaetarthria seminulum (Hbst.)	-	1	-	-	-	-
Hydraenidae						
Ochthebius lenensis Popp.	1	1	-	-	-	-
O. minimus (F.)	-	2	-	-	-	-
Ochthebius spp.	5	15	-	-	-	-
Hydraena britteni Joy/*riparia* Kug.	-	1	-	-	-	-
Ptiliidae						
Acrotrichis spp.	-	2	-	-	-	-
Leiodidae						
Catops sp.	-	-	-	-	-	1
Silphidae						
Thanatophilus sp.	1	1	-	1	-	-
Staphylinidae						
**Olophrum boreale* Payk.	-	1	3	-	-	-
O. fuscum (Grav.)	1	39	31	3	16	3
**O. rotundicolle* Sahl.	1	-	-	-	-	-
Arpedium brachypterum (Grav.)	-	48	4	4	31	1
Acidota crenata (F.)	-	2	-	-	1	-
A. cruentata Mann.	-	1	1	-	-	-
**Boreaphilus henningianus* Sahl.	-	2	-	-	1	-
**Pycnoglypta lurida* (Gyll.)	1	-	-	-	-	-
Bledius fuscipes Rye	-	-	5	4	-	1
Bledius spp.	-	-	2	15	-	1
Carpelimus corticinus (Grav.)	-	-	1	-	-	-
Platystethus nodifrons (Mann.)	-	2	-	-	-	-
Anotylus rugosus (F.)	-	4	-	-	-	-
Stenus juno (Payk.)	-	3	-	-	-	-
S. pallitarsus Steph.	-	1	-	-	-	-
Stenus spp.	-	14	2	3	10	1
Euaesthetus laeviusculus Mann.	-	1	-	-	-	-
Lathrobium impressum Heer	-	2	-	-	-	-
Lathrobium spp.	-	3	-	-	-	-
Ochthephilum fracticorne (Payk.)	-	1	-	-	-	-
Philonthus punctus (Grav.)/ **binotatus* (Grav.)	-	2	-	-	-	-
Philonthus sp.	-	1	-	-	-	-
Gabrius sp.	-	1	-	-	-	-
Quedius cruentus (Ol.)	-	1	-	-	-	-
Quedius sp.	1	-	-	-	-	-
Quedius/*Philonthus* spp.	-	4	-	-	1	-
Gymnusa brevicollis (Payk.)	-	-	-	-	1	-
G. variegata Keis.	-	1	-	-	-	-
Aleocharinae indet.	-	26	9	8	7	5
Scarabaeidae						
Aphodius merdarius (F.)	-	2	1	1	-	-
**A. montivagus* Erich.	-	3	-	-	1	-
**A. obscurus* F.	-	1	-	-	-	-
Aphodius spp.	3	7	2	2	-	-

cont...

Taxon	1	2	3	4	5	6
prob. *Cyphon* spp.	1	26	-	-	-	-
Byrrhidae						
**Simplocaria metallica* Sturm.	-	-	-	1	-	-
S. semistriata (F.)	-	1	-	-	-	-
Simplocaria sp.	-	-	-	1	-	-
Cytilus sericeus (Forst.)	-	1	-	-	-	-
Dryopidae						
Dryops sp.	-	1	-	-	-	-
Cantharidae						
Rhagonycha femoralis (Brul.)/ *lignosa* (Müll.)	-	1	-	-	-	
Lathridiidae						
Corticarina sp.	-	1	-	-	-	-
Scraptiidae						
Anaspis sp.	-	2	-	-	-	-
Chrysomelidae						
Plateumaris sericea (L.)	-	3	-	-	-	-
Phaedon sp.	-	1	-	-	-	-
Phyllodecta sp.	-	-	-	-	1	-
Apionidae						
Apion minimum Hbst.	-	2	-	-	-	-
Apion sp.	-	1	-	-	-	-
Curculionidae						
Otiorhynchus nodosus (Müll.)	-	-	1	-	-	-
O. ovatus (L.)	1	1	-	-	-	-
O. rugifrons (Gyll.)	-	1	-	-	-	-
Sitona suturalis Steph.	-	1	-	-	-	-
Bagous (?) *limosus* (Gyll.)	-	1	-	-	-	-
Bagous spp.	3	4	-	-	-	-
Notaris aethiops (F.)	-	20	-	1	-	-
N. bimaculatus (F.)	2	2	-	-	-	-
N. scirpi (F.)	1	-	-	-	-	-
Thryogenes sp.	-	2	-	-	-	-
Grypus equiseti (F.)	-	1	-	-	-	-
Phytobius canaliculatus Fah.	-	1	-	-	-	-
Limnobaris pilistriata (Steph.)	-	6	-	-	2	-

Sample locations:
1. silt beneath basal peat, Messingham.
2. basal peat, Messingham.
3. within Messingham Sands, Messingham.
4. top of Messingham Sands, Messingham.
5. basal peat, Flixborough Warren.
6. within Messingham Sands, Flixborough Warren.

* = species no longer recorded from Britain.

Taxonomy follows Kloet & Hincks (1964; 1977) and Freude *et al.* (1961-77).

cont...

Fig 8.9 cont...

Spider from Messingham Sands

Arachnida						
Araneae						
Thomisidae						
Oxyptila sp.	-	-	-	1	-	-

Taxonomy follows Locket & Millidge (1953)

ACKNOWLEDGEMENTS

The fieldwork and preliminary research for this paper were carried out during the tenure of an N.E.R.C. award in the Department of Geology, University of Birmingham during 1970-72. It owes much to discussion with colleagues in the University, particularly Professor F.W. Shotton, G.R. Coope, B.D. Giles and G.T. Warwick. Particle size analyses are the work of P. Foster and G. Coldicott typed the final manuscript. The final draft benefitted considerably from the editorial ability of J.A. Catt and Gudrún Sveinbjarnardóttir. The figures are the work of J. and G. Dowling of the Department of Geography. Access to sites was kindly arranged by the successive managers of British Industrial Sands' quarries at Messingham and by M. Elford of the British Steel Corporation. Discussion in the field with T. Fletcher, G.D. Gaunt, E.A. Koster, A.B. Sumpter and J.A. Taylor is gratefully recorded and the finding of the flint scraper by my wife is also noted. Thanks are also due to J. Baxter, the late M. Dolby, M.A. Girling, J.R.A. Greig, C.A. Howes, C. Johnson, N. Loughlin, J. Rackham, G. Scherer, P. Skidmore and R.T. Thompson, amongst many others. Access to insect collections housed in the British Museum, Doncaster Museum and Manchester Museum is also acknowledged.

BIBLIOGRAPHY

1. Angus, R.B., 1973, Pleistocene *Helophorus* (Coleoptera, Hydrophilidae) from Borislav and Starunia in the Western Ukraine, with a reinterpretation of M. Komnicki's species, description of a new Siberian species, and comparison with British Weichselian faunas, *Philosophical Transactions of the Royal Society of London B265*, 299-326.

2. Armstrong, A.L., 1931, A late Upper Aurignacian station in north Lincolnshire, *Proceedings of the Prehistoric Society of East Anglia, 6*, 335-339.

3. Armstrong, A.L., 1932a, Upper Palaeolithic and Mesolithic Stations in N. Lincolnshire, *Proceedings of the Prehistoric Society of East Anglia, 7*, 130-131.

4. Armstrong, A.L., 1932b, An Upper Aurignacian station on the Lincolnshire Cliff, *Proceedings of the International Congress of Prehistoric and Protohistoric Science*, 74-75, (London).

5. Armstrong, A.L., 1956, Prehistory - Palaeolithic, Neolithic and Bronze Ages, in: *Sheffield and its Region*, ed. Linton, D.L., (British Association for the Advancement of Science, Sheffield), 90-110.

6. Ashworth, A.C., 1972, A Late Glacial insect fauna from Red Moss, Lancashire, England, *Entomologica Scandinavia, 3*, 211-224.

7. Ashworth, A.C., 1973, The climatic significance of a late Quaternary insect fauna from Rodbaston Hall, Staffordshire, *Entomologica Scandinavia, 4*, 191-205.

8. Balfour-Browne, F., 1950, 1958, *British Water Beetles*, (Ray Society, London), vols. II & III.

9. Balthazar, V., 1964, *Monographie der Scarabaeidae und Aphodiidae der Palaearktischen und Orientalischen Region*, (Prague).

10. Bartley, D.D., 1962, The Stratigraphy and Pollen Analysis of Lake Deposits near Tadcaster, Yorkshire, *New Phytologist, 61*, 277-287.

11. Bellamy, D.J., 1972, Templates of Peat Formation, *Proceedings of the 4th. International Peat Congress, I*, (Otaniemi), 7-17.

12. Bishop, W.W. and Coope, G.R., 1977, Stratigraphical and Faunal Evidence for Lateglacial and Early Flandrian Environments in South-West Scotland, in: *Studies in the Scottish Lateglacial Environment*, ed. Gray, J.M. and Lowe, J.J., (Pergamon, Oxford), 61-88.

13. Black, R.F., 1951, Eolian deposits of Alaska, *Arctic, 4*, 89-111.

14. Brooks, C.E.P., 1926, *Climate through the Ages*, (Benn, London).

15. Brown, R.J.E., 1970, Permafrost as an ecological factor in the Subarctic, *Proceedings of the Helsinki Symposium on the Ecology of the Subarctic regions*, (UNESCO, Paris), 129-140.

16. Buckland, P.C., 1976, The use of insect remains in the interpretation of archaeological environments, *Geo-archaeology*, ed. Davidson, D.A. and Shackley, M.L., (Duckworth, London), 369-396.

17. Buckland, P.C., 1979, *Thorne Moors: a palaeoecological study of a Bronze Age site*, Department of Geography, University of Birmingham, occasional publication, 8.

18. Buckland, P.C., 1981, *A Bibliography of Quaternary Entomology*, Department of Geography, University of Birmingham, working paper series, 11.

19. Buckland, P.C., Coope, G.R. and Gaunt, G.D., in prep., Late Glacial stratigraphy and insect faunas on the Hatfield Levels.

20. Buckland, P.C. and Dolby, M.J., 1973, Mesolithic and Later Material from Misterton Carr, Notts. - an interim report, *Transactions of the Thoroton Society of Nottinghamshire, 77*, 5-33.

21. Campbell, J.B., 1977, *The Upper Palaeolithic of Britain*, (University Press, Oxford).

22. Coiffait, H., 1967, Tableau de Determination des *Philonthus* de la Region Palearctique Occidentale (Col., Staphylinidae), *Annales de la Societe Entomologique de France, n.s., 3*, 381-450.

23. Coope, G.R., 1965, Fossil Insect Faunas from Late Quaternary Deposits in Britain, *Advancement of Science*, March 1965, 564-575.

24. Coope, G.R., 1966, *Diachila* (Col., Carabidae) from the Glacial Deposits at Barnwell Station, Cambridge, *Entomologist's monthly Magazine*, *102*, 119-120.

25. Coope, G.R., 1973, Tibetan Species of Dung Beetle from Late Pleistocene Deposits in England, *Nature*, *245*, 335-336.

26. Coope, G.R., 1975, Climatic fluctuations in northwest Europe since the Last Interglacial, indicated by fossil assemblages of Coleoptera, in: *Ice Ages: Ancient and Modern*, ed. Wright, A.E. and Moseley, F., (Seel House, Liverpool), 153-168.

27. Coope, G.R., 1977, Fossil Coleoptera assemblages as sensitive indicators of climatic changes during the Devensian (Last) cold stage, *Philosophical Transactions of the Royal Society of London*, *B280*, 313-340.

28. Coope, G.R. and Brophy, J.A., 1972, Late Glacial environmental changes indicated by a coleopteran succession from North Wales, *Boreas*, *1*, 97-142.

29. Coope, G.R. and Joachim, M., 1980, Lateglacial environmental changes interpreted from fossil Coleoptera from St. Bees, Cumbria, N.W. England, in: *Studies in the Lateglacial of North-West Europe*, ed. Lowe, J.J., Gray, J.M. and Robinson, J.E., (Pergamon, Oxford), 55-68.

30. Coope, G.R. and Osborne, P.J., 1968, Report on the coleopterous fauna of the Roman Well at Barnsley Park, Gloucestershire, *Transactions of the Bristol and Gloucestershire Archaeological Society*, *86*, 84-87.

31. Coope, G.R. and Sands, C.H.S., 1966, Insect faunas of the last glaciation from the Tame Valley, Warwickshire, *Proceedings of the Royal Society, of London*, *B244*, 389-412.

32. Coope, G.R., Shotton, F.W. and Strachan, I., 1961, A Late Pleistocene Fauna and Flora from Upton Warren, Worcestershire, *Philosophical Transactions of the Royal Society of London*, *B244*, 397-412.

33. Dakyns, J.R., Fox-Strangeways, C. and Cameron, A.G., 1886, *The Geology of the Country between York and Hull*, (HMSO, London).

34. Downes, J.A., 1965, Adaptations of Insects in the Arctic, *Annual Review of Entomology*, *10*, 257-274.

35. Dudley, H.E., 1931, *The History and Antiquities of Scunthorpe and Frodingham District*, (Scunthorpe).

36. Dudley, H.E., 1949, *Early Days in North-West Lincolnshire*, (Scunthorpe).

37. Edwards, W., 1936, Pleistocene dreikanter in the Vale of York, *Memoirs of the Geological Survey of Great Britain, Summaries of Progress for 1934, pt.2*, 8-19.

38. Edwards, W., 1967, *Geology of the Country around Ollerton*, (HMSO, London).

39. Embleton, C. and King, C.A.M., 1975, *Periglacial Geomorphology*, (Arnold, London).

40. Evans, J.G., 1975, *The Environment of Early Man in the British Isles*, (Elek, London).

41. Evans, W.B., Wilson, A.A., Taylor, B.J. and Price, D., 1968, *Geology of the Country around Macclesfield, Congleton, Crewe and Middlewich*, (HMSO, London).

42. Fitter, A., 1978, *An Atlas of the Wild Flowers of Britain and Northern Europe*, (Collins, London).

43. Freude, H., 1976, Familie Carabidae, in Freude *et al.*, II.

44. Freude, H., Harde, K.W. and Lohse, G.A., 1961-1977, *Die Käfer Mitteleuropas*, (Goecke and Evers, Krefeld).

45. Gaunt, G.D., 1976, The Devensian Maximum Ice Limit in the Vale of York, *Proceedings of the Yorkshire Geological Society*, *40*, 631-637.

46. Gaunt, G.D., 1981, Quaternary History of the Southern Part of the Vale of York, in: *The Quaternary in Britain*, ed. Neale, J. and Flenley, J., (Pergamon, Oxford), 82-97.

47. Gaunt, G.D., Jarvis, R.A. and Matthews, B., 1971, The Late Weichselian Sequence in the Vale of York, *Proceedings of the Yorkshire Geological Society*, *38*, 281-284.

48. Godwin, H., 1975. *The History of the British Flora*, 2nd. ed., (Cambridge, London).

49. Grigson, C., 1978, The Late Glacial and Early Flandrian Ungulates of England and Wales - an interim review, in: *The effect of man on the Landscape: the Lowland Zone*, (Council for British Archaeology, London), 46-56.

50. Harrison, R.K., 1968, Petrography of Some Quaternary Sands, in Evans *et al.*, 253-258.

51. Hansen, V., Klefbeck, E., Sjöberg, O., Stenius, G. and Strand, A., 1960, *Catalogus Coleopterorum Fennoscandiae et Daniae*, revised Lindroth, C.H., (Lund).

52. Hoffman, A., 1954, 1958, Coleopteres Curculionidae, *Faune de France*, *59*, 62.

53. Holland, S.M., 1975, Pollen analytical investigations at Crosby Warren, Lincolnshire in the vicinity of the Iron Age and Romano-British settlement of Dragonby, *Journal of Archaeological Science*, *2*, 353-364.

54. Horion, A., 1951, *Verzeichnis der Käfer Mitteleuropas*, (Stuttgart).

55. Jelgersma, S., 1966, Sea level changes during the last 10 000 years, in: *World Climate from 8000 to 0 B.C.*, (Royal Meteorological Society, London).

56. Jones, R.L. and Gaunt, G.D., 1976, A Dated Late Devensian organic deposit at Cawood, near Selby, *Naturalist, 101*, 121-123.

57. Jong, J.D. de, 1967, The Quaternary of the Netherlands, in: *The Quaternary*, ed. Rankama, K., (Interscience, London), 2, 301-426.

58. Joy, N.H., 1932, *A Practical Handbook of British Beetles*, (Witherby, London).

59. Kendall, P.F. and Wroot, H.E., 1924, *Geology of Yorkshire*, (Vienna).

60. Kloet, G.S. and Hincks, W.D., 1964, *A Checklist of British Insects*, 2 Hemiptera and small orders., 2nd. ed., (Royal Entomological Society, London).

61. Kloet, G.S. and Hincks, W.D., 1977, *A Checklist of British Insects*, 4 Coleoptera and Strepsiptera, 2nd. ed. rev. Pope, W.D. (Royal Entomological Society, London).

62. Lacaille, A.D., 1946, Some Flint Implements of special interest from Lincolnshire, Hampshire and Middlesex, *Antiquaries Journal, 26*, 180-185.

63. Landin, B.-O., 1961, Ecological Studies on Dung Beetles, *Opuscula Entomologica*, suppl. 19.

64. Larsson, S.G. and Gigja, G., 1959, Coleoptera, in: *The Zoology of Iceland, III, 46* (Munksgaard, Copenhagen).

65. Lindroth, C.H., 1935, The Boreo-British Coleoptera, *Zoogeographica, 2*, 579-634.

66. Lindroth, C.H., 1940, Zur Systematik fennoskandischen Carabiden, 4-12, *Bembidion*-Studien, *Notulae entomologicae, 19*, 63-99.

67. Lindroth, C.H., 1945, 1949, *Die Fennoskandischen Carabidae : Eine Tier-geographische Studie*, Meddelanden fran Goteborgs Musei Zoologiska Avdelsing.

68. Lindroth, C.H., 1961, The Ground-Beetles of Canada and Alaska (Carabidae excl. Cicindelinae), *Opuscula entomologica, suppl. 20*.

69. Lindroth, C.H., 1974, Carabidae, *Handbooks for the Identification of British Insects, 4, 2*, (Royal Entomological Society, London(.

70. Locket, G.H. and Millidge, A.F., 1953, *British Spiders*, (Ray Society, London).

71. Lohse, G.A., 1964, Familie Staphylinoidea 1 (Micropeplinae bis Tachyporinae), in Freude *et al., 4*.

72. Loughlin, N., 1977, Dales Ware: A Contribution to the Study of Roman Coarse Pottery, in: *Pottery and Early Commerce*, ed Peacock, D.P.S., (Academic Press, London), 85-146.

73. Lowe, J.J. and Gray, J.M., 1980, The stratigraphic subdivision of the Lateglacial of NW Europe: a discussion, in: *Studies in the Lateglacial of North-West Europe*, ed. Lowe, J.J., Gray, J.M. and Robinson, J.E. (Pergamon, Oxford), 157-177.

74. Maarleveld, G.C., 1960, Wind directions and coversands in the Netherlands, *Biuletyn Peryglacyn, 8,* 49-58.

75. Machatschke, J.W. 1969, Lamellicornia, in Freude *et al., 8,* 265-371.

76. McKee, E.D., 1964, Inorganic Sedimentary Structures, in: *Approaches to Palaeoecology,* ed. Imbrie, J. and Newell, N., (Wiley, London).

77. Matthews, B., 1970, Age and Origin of Aeolian Sands in the Vale of York, *Nature, 227,* 1234-1236.

78. May, J., 1976, *Prehistoric Lincolnshire,* (Lincoln).

79. Mellars, P.A., 1974, The Palaeolithic and Mesolithic, in: *British Prehistory, a new outline,* ed. Renfrew, C., (Duckworth, London), 41-99.

80. Mohr, K.-H., 1966, Familie Chrysomelidae, in Freude *et al., 9,* 95-280.

81. Moore, P.D. and Bellamy, D.J., 1973, *Peatlands,* (Elek, London).

82. Morgan, A., 1973, Late Pleistocene environmental changes indicated by fossil insect faunas of the English Midlands, *Boreas, 2,* 173-212.

83. Nicolas, J.-L., 1971, 3e Contribution a l'Etude des Aphodiini de la Faune Francaise: Le Complex *A. (Aegolius) mixtus* Villa (Col., Aphodiidae). *Bulletin Mensuel de la Societe Linné de Lyon, 40,* 154-160.

84. Osborne, P.J., 1972, Insect Faunas of Late Devensian and Flandrian age from Church Stretton, Shropshire, *Philosophical Transactions of the Royal Society of London, B263,* 327-367.

85. Osborne, P.J., 1973, A Late-Glacial Insect Fauna from Lea Marston, Warwickshire, *Proceedings of the Coventry and District Natural History and Scientific Society, 4,* 209-213.

86. Osborne, P.J., 1974, An Insect Assemblage of Early Flandrian Age from Lea Marston, Warwickshire and its bearing on the Contemporary Climate and Ecology, *Quaternary Research, 4,* 471-486.

87. Osborne, P.J., 1980, The Late Devensian - Flandrian transition depicted by serial insect faunas from West Bromwich, Staffordshire, England, *Boreas, 9,* 139-147.

88. Paddaya, K., 1972, The Late Palaeolithic of the Netherlands - a review, *Helinium, 11,* 257-270.

89. Palm, T., 1952, Om tva for Sverige nya *Philonthus*-arter (Col., Staphylinidae), *Opuscula entomologica, 17,* 117-124.

90. Paulian, R., 1959, Coleopteres Scarabeides, *Faune de France,* 63.

91. Pearson, R.G., 1961, The Coleoptera from a Detritus Mud Deposit of Full Glacial Age at Colney Heath, near St. Albans, *Proceedings of the Linnean Society of London, 173*, 37-55.

92. Pennington, W., 1975, A chronostratigraphic comparison of Late Weichselian and Late-Devensian subdivisions, illustrated by two radiocarbon-dated profiles from western Britain, *Boreas, 4*, 157-171.

93. Pennington, W. and Lishman, J.P., 1971, Iodine in Lake Sediments in Northern England and Scotland, *Biological Review, 46*, 279-313.

94. Penny, L.F., Coope, G.R. and Catt, J.A., 1969, Age and insect fauna of the Dimlington Silts, East Yorkshire, *Nature, 224*, 65-67.

95. Pettijohn, F.J., 1957, *Sedimentary Rocks*, (Harper, New York).

96. Péwé, T.L., 1969, The Periglacial Environment, in: *The Periglacial Environment*, ed Péwé, T.L., (McGill-Queen's U.P., Montreal), 1-10.

97. Pryme, A. de la, 1695, *Ephemeris Vitae Abrahami Pryme*, ed. Jackson, C., Publications of the Surtees Society, 54, 1869.

98. Radley, J., 1969, A note on four Maglemosian bone points from Brandesburton and a flint site at Brigham, Yorkshire, *Antiquaries Journal, 49*, 377-378.

99. Radley, J. and Simms, C., 1967, Wind erosion in east Yorkshire, *Nature, 216*, 20-22.

100. Reitter, E., 1909, *Fauna Germanica - Die Käfer des Deutches Reiches*, Stuttgart.

101. Richert, D.E. and Tedrow, J.C.F., 1967, Pedological Investigations on some Aeolian Deposits of Northern Alaska, *Soil Science, 104*, 250-262.

102. Ruddiman, W.F. and McIntyre, A., 1981, The North Atlantic Ocean during the last deglaciation, *Palaeogeography, Palaeoclimatology, Palaeoecology, 35*, 121-144.

103. Rutten, M.G., 1954, Deposition of Coversands and Loess in the Netherlands, *Geologie en Mijnbouw, 16*, 127-129.

104. Shotton, F.W., 1972, An Example of Hard-Water Error in Radiocarbon Dating of Vegetable Matter, *Nature, 240*, 460-461.

105. Shotton, F.W. and Osborne, P.J., 1965, The Fauna of the Hoxnian Interglacial Deposits of Nechells, Birmingham, *Philosophical Transactions of the Royal Society of London, B248*, 353-378.

106. Sissons, J.B., 1980, Palaeoclimatic inferences from Loch Lomond Advance glaciers, in: *Studies in the Lateglacial of North-West Europe*, ed. Lowe, J.J., Gray, J.M. and Robinson, J.E., (Pergamon, Oxford), 31-44.

107. Smithson, F., 1931, The Triassic Sandstones of Yorkshire and Durham, *Proceedings of the Geologists' Association, 42,* 125-154.

108. Stainforth, T., 1944, Reed beetles of the Genus *Donacia* and its allies in Yorkshire, *Naturalist, 810,* 81-91; *811,* 127-139.

109. Straw, A., 1963, Some Observations on the Cover Sands of North Lincolnshire, *Transactions of the Lincolnshire Naturalists' Union, 15,* 260-269.

110. Suggate, R.P. and West R.G., 1959, The extent of the last glaciation in eastern England, *Proceedings of the Royal Society of London, B150,* 263-283.

111. Swinnerton, H.H., 1914, Periods of dreikanter formation in South Notts., *Geological Magazine, 6,* 208-211.

112. Thompson, H.A., 1968, Climate, in: *Science, History and Hudson Bay,* ed. Beals, C.S. and Shenstone, D.A., (Ottawa), *1,* 263-286.

113. Ussher, W.A.E., 1890, *The Geology of Parts of North Lincolnshire and South Yorkshire,* (HMSO, London).

114. Waterbolk, H.T., 1954, *De Praehistorische Mens en Zijn Mulieu,* (Assen).

115. Williams, R.B.G., 1975, The British Climate during the Last Glaciation; an interpretation based on periglacial phenomena, in: *Ice Ages: Ancient and Modern,* ed. Wright, A.E. and Moseley, F., (Seel House, Liverpool).

116. Wilson, P., Bateman, R.M. and Catt, J.A., 1981, Petrography, origin and environment of deposition of the Shirdley Hill Sand of southwest Lancashire, *Proceedings of the Geologists' Association, 92,* 211-229.

117. Wilson, V., 1948, *East Yorkshire and Lincolnshire,* British Regional Geology (HMSO).

118. Wright, R.F. and Heinselman, M.L., 1973, The Ecological Role of Fire in Natural Conifer Forests of Western and Northern America, *Quaternary Research, 3,* 317-328.

119. Zeuner, F., 1959, *The Pleistocene Period,* (Hutchinson, London).

9. Medial moraines on valley glaciers

R.J. Small

1. INTRODUCTION

Medial moraines, one of the best known features of valley glaciers, are frequently regarded as originating from the junction of two ice-streams, the lateral moraines of which amalgamate to form the medial moraine. The latter may comprise a debris ridge stretching for several km down the trunk glacier, to terminate at the snout or merge into the general spread of ablation moraine that covers the lowermost parts of 'dirty' glaciers. Such moraines have been referred to as of the Ice-stream Interaction Type (ISI) by Eyles and Rogerson (1978).

Detailed study of medial moraines in the field reveals, however, that this simple explanation is inadequate to account for what are usually complex landforms. In the first place, identification of the actual source of the debris 'feeding' the moraine may be highly problematical. The angular nature of this material initially suggests a predominantly supraglacial origin, via congelifraction of exposed peaks and slopes and direct gravity fall onto the surfaces of the contributing ice-streams. However, Sharp (1949) has noted that medial moraines frequently mark the outcrops of 'steep, debris-rich zones in glaciers'; he adds that 'the debris comprising such moraines is largely a residual accumulation of englacial debris left on the surface by melting'. In other words only a small proportion of the debris cover of the medial moraine 'has always been superglacial'. Sharp, in his study of the constitution of valley glaciers (1948), proposes that these debris-rich zones ('septa' or 'longitudinal septa') develop not only at the junctions of 'juxtaposed' ice-streams but also where very small tributary glaciers join major glaciers to form 'superimposed' and 'inset' ice-streams.

Another problem relates to the considerable morphological variability of medial moraines. There is an immediate contrast between uninterrupted moraine ridges such as those of the Aletsch and Gorner glaciers in the

Swiss Alps and irregular 'beaded' moraines such as those of Austerdalsbreen, Norway, of which Embleton and King (1968) write: 'the beading appears to have an annual spacing, and can be accounted for by the more active production of waste during the summer season'. Since this beading evidently relates to englacial debris being revealed at the glacier surface by summer ablation, there is the additional problem of explaining the mechanism of debris incorporation, presumably on the accumulation zone of the glacier or (in the case of Austerdalsbreen, which is characterised by a major ice-fall) adjacent to or beneath the attenuated, highly fractured ice of the ice-fall. Whatever the mechanism of incorporation, it is evident that the Austerdalsbreen moraine is of the Ablation Dominant Type (AD) postulated by Eyles and Rogerson (1978). The debris forming such moraines is 'held englacially and revealed downglacier by ablation'. Two sub-types of ablation dominant moraines are recognised: the Above Firn-line Sub-type, in which debris is incorporated on the accumulation zone by way of burial by winter snow, and the Below-Firn-line Sub-type, in which the debris enters the glacier by crevasses on the ablation zone but is rapidly re-exposed at the glacier surface by summer melting. Although Eyles and Rogerson make a firm distinction between Ice-stream Interaction and Ablation Dominant moraines, it is clear from Sharp's observations that a composite type may frequently occur, where debris falling onto the margins of ice-streams within the accumulation zone is incorporated rather than forming surface lateral moraines. Amalgamation of ice-streams then produces longitudinal englacial septa which are revealed by ablation below the firn-line to form medial moraine ridges.

The possible effects of major ice structures on medial moraine development need to be commented on. It is noticeable that in some instances medial moraines, usually of beaded form, 'grow out' of glacier tongues at the foot of ice-falls. The two medial moraines of the Glacier de Tsidjiore Nouve (Small and Clark 1974) display this relationship. Posamentier (1978), in a discussion of surface 'waves' and ogives commonly developed at the base of ice-falls, suggests that decreased ice velocity related to sharply reduced bed gradient results in strongly compressive flow (contrasting with the high velocity extending flow of the ice-fall). This may in turn produce either strong isoclinal folding, with axial plane reverse faulting, or large-scale reverse faulting, with associated drag folding, within the glacier in the zone at the base of the ice-fall. In either instance there is the strong possibility of debris-rich folia, developed at the base of the glacier as it passes through the ice-fall being brought directly to, or close to, the glacier surface. Such subglacial debris would tend to outcrop as bands or zones running transversely across the glacier, producing beaded moraine structures such as have been described. Whether or not ice-falls promote medial moraine formation, there is certainly evidence to show that they do not destroy moraines. Sharp (1949), in his study of Wolf Creek Glacier, notes that the debris constituting a medial moraine above the ice-fall collects within the many deep crevasses developed on the ice-fall, and is then revealed again at the glacier surface by ablation at the base of the ice-fall. 'Part of the debris then spills back over the unmantled ice, and surficial continuity of the moraine is re-established. The only noticeable change is a somewhat greater width produced by lateral spreading'.

2. STUDIES OF MEDIAL MORAINE FORM AND GENESIS

(1) The Kaskawulsh Glacier, Yukon

This is a 'classic' ISI type of medial moraine, and has been investigated in detail by Loomis (1970). However, morphologically it shows some complexity. It comprises a number of individual longitudinal ridges, each of which is ice-cored, with a surface debris cover usually only 5-10 cm in thickness. The occurrence of the ice core indicates that the major process determining moraine relief is differential ablation (the contrast in ablation between bare ice and debris covered 'protected' ice). The moraine complex as a whole undergoes systematic changes of form downglacier. At the point of confluence of contributing ice-streams it is 1 km in width; the debris then narrows rapidly to 60 m at a distance of only 0.75 km downglacier; this reduced width then remains approximately constant for several kilometres. In terms of height, the moraine ridge at the confluence is weak (less than 2 m), but grows downglacier to 20 m at a distance of 1.4 km; it then declines to 7 m after 5 km, thereafter remaining approximately constant in height for several more kilometres until the snout.

It is, therefore, possible to identify WAXING, WANING and CONSTANT stages of development for the Kaskawulsh moraine (the terminology is that of the present author, not that of Loomis). The factors influencing these stages are identified by Loomis as follows:

(i) Differential ablation (as stated above). By measurement of 31 ablation stakes Loomis demonstrated an inverse relationship between mean daily ablation and debris thickness (correlation coefficient -0.91). However, a very thin debris layer (less than 1 cm) produced an accelerated rate of ablation in relation to that of bare ice.

(ii) Lateral ice compression below the confluence of contributing ice-streams. This effectively increased debris thickness, reduced ablation, and promoted increase in moraine height.

(iii) Extending flow. This operates on the Kaskawulsh glacier from 1.5 km below the confluence, further reducing differential ablation and, in the view of Loomis, causing moraine relief to decline (hence the waning stage). Loomis argues that 'the till becomes thinner and less effective as an insulator, and ablation rates increase'. However, this is an inadequate explanation, since unless ablation at the moraine crest is actually greater than that of bare ice,

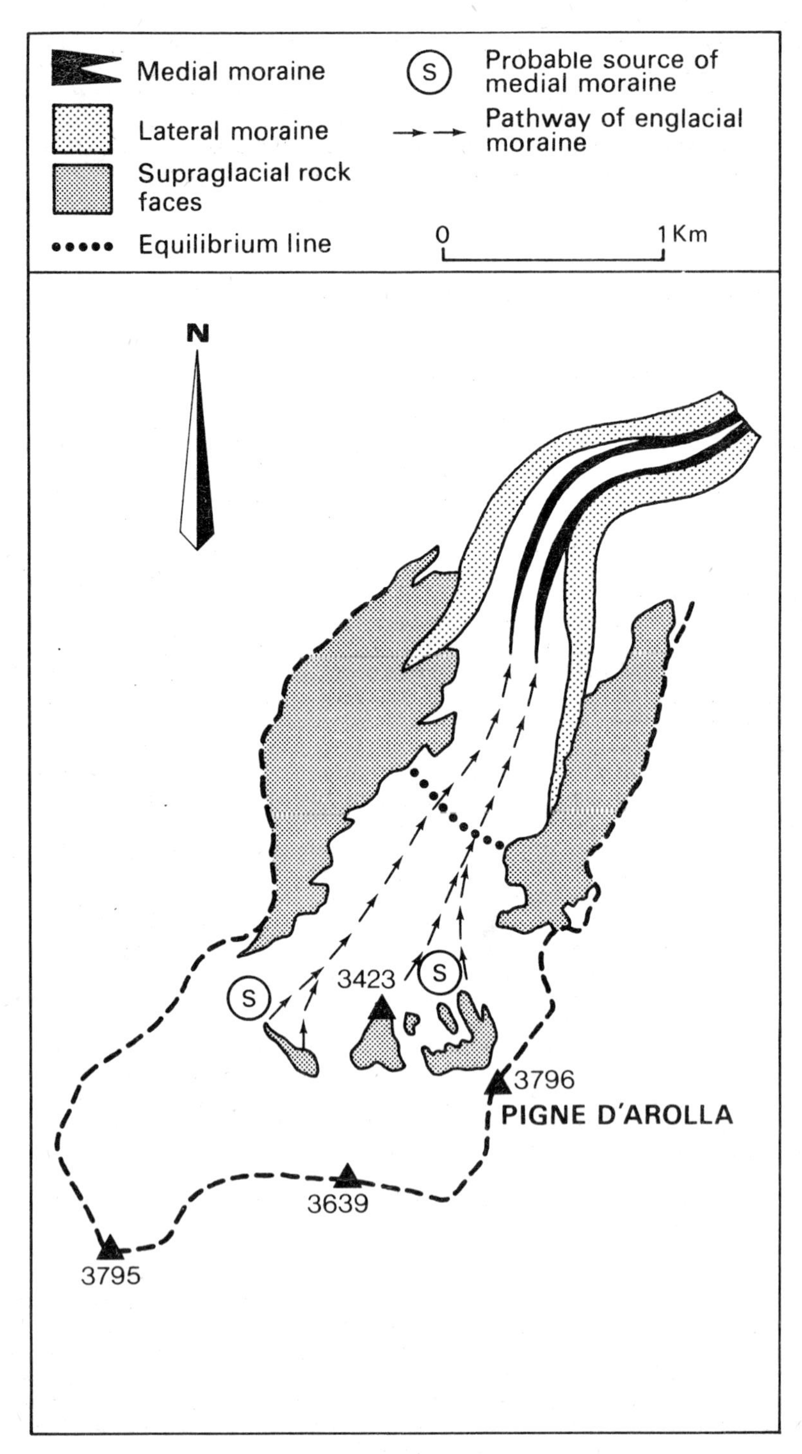

Figure 9.1. Map of the Glacier de Tsidjiore Nouve.

the moraine ridge will continue to grow, albeit more slowly. The constant section of the moraine poses further difficulties. Loomis's view that 'at a distance of approximately 5 km downglacier the moraine mantle has shifted to create a uniformly thinner layer', producing an 'equilibrium state' resulting in constant moraine relief, offers no real solution.

(2) The glaciers at Arolla, Valais, Switzerland

(a) The Glacier de Tsidjiore Nouve (figure 9.1)

This glacier possesses two medial moraines, which are not simple ISI moraines but 'emerge' as AD types some 400 m from the base of the Pigne d'Arolla ice-fall, which links the firn basin and the lower glacier tongue (Small and Clark, 1974). The larger southern (main) moraine attained a maximum height above adjacent bare ice of 22.25 m in 1971; by 1976 this had risen to 30.8 m. The maximum measured width was 101 m (1971). The smaller northern (subsidiary) moraine reached a maximum height of 6 m and a maximum width of 59 m in the 1971 survey. The two moraines are approximately 1 km in length (and are thus on a totally different scale from the Kaskawulsh moraine), extending to the glacier snout and merging with each other and extensive lateral moraines to give a continuous debris cover on the lowermost part of the glacier tongue.

Other morphological features of note are as follows:

(i) When fully developed both moraines constitute single longitudinal ice-cored ridges, with a debris cover of only 5-6 cm in thickness extending between individual boulders and other larger clasts.

(ii) Both moraines begin to form as a series of discontinuous mounds and ridges, the latter orientated transversely to the glacier axis (figure 9.2). Eyles (in correspondence, Small and Clark, 1976) has suggested that the concentration of surface debris needed to produce this beaded effect may be a 'secondary' process. An initially even spread of supraglacial debris will become discontinuous as a result of sliding into shallow transverse troughs produced by the more rapid ablation of 'dark' ogives. When the debris concentrations are sufficient to retard ablation, a form of relief inversion occurs, with dark ogives giving ice-cored 'debris ridges' and white ogives intervening debris-free 'corridors'. However, a detailed field examination (see below) has shown that all the mounds and ridges of the Glacier de Tsidjiore Nouve are coincident with clearly defined englacial debris bands.

(iii) Both moraines grow downglacier to a maximum height, and then undergo decline as the debris cover becomes attenuated, primarily by lateral sliding. However, as noted above, the ridges do continue in subdued form as far as the glacier snout; in broad terms these lowermost sections may be described as constant in height. Thus, like the Kaskawulsh moraine, the Tsidjiore Nouve moraines exhibit waxing, waning and constant stages.

Figure 9.2. View of the Glacier de Tsidjiore Nouve, looking towards the base of the Pigne d'Arolla ice-fall. Note the 'emergence' of the debris and the 'beaded' form.

Detailed field study of the Tsidjiore Nouve moraines has shown that their form is the product of several inter-related factors and processes (figure 9.3). The strong causal links between debris supply (assumed to be initially from supraglacial or subglacial sources, but transmitted by way of englacial pathways), the increasing thickness of surface till as debris is melted out downglacier, differential ablation (defined here as $Da = \frac{\text{clean ice ablation}}{\text{covered ice ablation}}$, giving a value normally >1.0) and developing moraine relief will be readily apparent. In the zone at the base of the ice-fall strong compressive flow (and possibly some lateral ice compression, resulting from the fact that the glacier tongue is confined by massive stable banks of lateral moraine) also contributes to increased thickness of the debris cover. The very important role of differential ablation can be demonstrated by ablation transects across the glacier surface. Information about the englacial debris content of the glacier in the zone at the base of the ice-fall is not readily available, though an investigation by Grande Dixence S.A., using thermal-electric probes designed to penetrate to the subglacial floor,

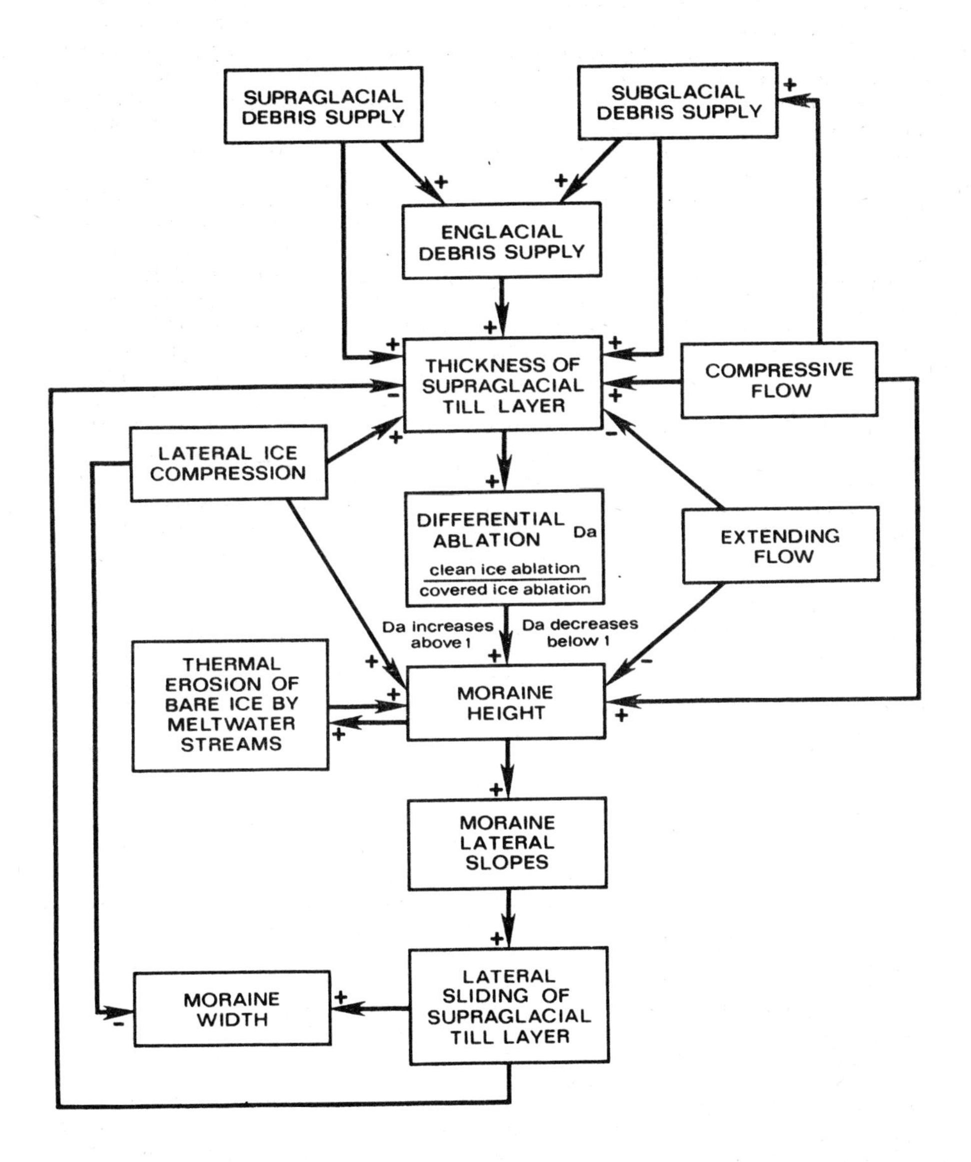

Figure 9.3. The factors and processes of medial moraine growth and decline.

yielded possibly significant results. Of the 40 probes inserted between the ice-fall and the heads of the medial moraines, 22 reached glacier bottom at depths of 160-183 m but 18 failed, presumably as a result of encountering englacial clasts. Of the latter, half failed at depths of 48-95 m, and none below 110 m, suggesting the possibility of broad septa of englacial debris largely confined to the uppermost half of the glacier, and also supporting the view that the englacial material is supraglacially derived.

Figure 9.3 also demonstrates the positive relationships between moraine height, moraine slopes, debris sliding and moraine width. Surveyed cross-profiles of the main Tsidjiore Nouve moraine show that within a few hundred metres of the 'source' maximum lateral slopes attain 27-29°, evidently the repose angle of the coarse platy debris comprising the greater part of the detrital cover. Any further increase in moraine height, owing to differential ablation and surface 'erosion' of adjacent bare ice by concentrated meltwater streams, accelerates lateral sliding (particularly of the larger clasts which, because of greater momentum, gravitate rapidly to the base of the moraine slopes) and thus results in steadily growing moraine width; on the Tsidjiore Nouve main moraine width increases from 38 m to 101 m in a downglacier distance of 200 m from the point at which 29° lateral slopes are first attained.

Eventually this process of active lateral sliding (allied perhaps to a decrease in englacial debris supply downglacier as septa are consumed) necessarily produces some attenuation of the debris cover of the moraine ridge. The process may be assisted on the Glacier de Tsidjiore Nouve by extending flow, which evidently affects the middle section of the glacier tongue (it is strongly convex in long-profile), producing a series of major transverse crevasses on the ridge crest. Figure 9.3 indicates the possibility of debris attenuation to a critically low threshold, which induces a Da value of <1.0 (i.e. 'covered' ice melts more rapidly than adjacent bare ice). Some such mechanism is essential for the initiation of waning moraine relief. The effects of very thin debris layers in producing abnormally high ablation rates have been noted by several authors. Sharp (1949) suggested that where a 'one-particle' layer, through its capacity to absorb solar radiation and transmit heat energy to subjacent ice by conduction or re-radiation, has this effect the process should be termed '<u>indirect ablation</u>'. Østrem (1959) and Loomis (1970), as noted above, have demonstrated that such <u>accelerated ablation</u> (the term prefered by the present author for this phenomenon) is produced by debris layers of 0.5-1.0 cm thickness. Observations on the Glacier de Tsidjiore Nouve have strongly supported the important role of accelerated ablation. Figure 9.4 illustrates the results of an ablation transect taken across the peak of the main moraine and the waning section of the subsidiary moraine, where debris attenuation has resulted in thicknesses of only 1-2 cm. Note that the lowest ablation rates (1.6-2.1 cm/day) are found on the main moraine, beneath a cover of 5-6 cm, and that the highest ablation rates (4.3-4.6 cm/day) are on the flanks of the subsidiary moraine.

Site	Till thickness	Total ablation	Days of observation	Mean diurnal ablation
S. face main moraine	6 cm	64.3 cm	39	1.6 cm
Crest of main moraine	5 cm	52.8 cm	39	1.4 cm
N. face main moraine	6 cm	67.4 cm	32	2.1 cm
Clean ice	-	142.4 cm	36	3.9 cm
S. face subsid. moraine	2 cm	169.9 cm	39	4.3 cm
N. face subsid. moraine	1 cm	161.1 cm	35	4.6 cm
Clean ice	-	141.0 cm	36	3.9 cm

Figure 9.4. Tsidjiore Nouve Glacier: ablation transect. (24.7.1976 to 2.9.1976). Transect across peak of main medial moraine and declining section of subsidiary moraine.

The waxing and waning stages of medial moraine formation can, therefore, be satisfactorily explained in terms of differential ablation and accelerated ablation, related to systematic variations in debris thickness downglacier produced by the melting out of englacial debris and subsequent attenuation of the resultant layer by sliding and other mechanisms. The problem of the final constant stage remains unsolved. Observations on the Glacier de Tsidjiore Nouve indicate that this may represent a 'deceleration' of the waning 'mechanisms', induced by the following three processes. First, the onset of compressive flow towards the glacier snout causes some thickening of the debris cover and thus differential rather than accelerated ablation. Secondly, the merging of the two medial moraines with the extensive lateral moraines inhibits lateral sliding, with a similar effect. Thirdly - and perhaps most importantly - there is considerable 'recycling' of debris, some of which falls into transverse crevasses on the middle section of the moraine and re-emerges from its shallow englacial situation a few hundred metres downglacier; individual debris particles may be recycled in this way several times.

(b) The Bas Glacier d'Arolla.

This glacier is characterised by large lateral moraines and two smaller medial moraines which emerge quite close to the glacier snout. The latter both display the waxing stage of development; however, the waning stage is poorly represented, and is influenced less by lateral debris sliding than by slumping towards and down the steep terminal face of the glacier. Detailed features of the two moraines are as follows:

(i) The western moraine begins as a patchy cover of boulders and smaller clasts, which merge downglacier to give a continuous strip of debris. The moraine ridge proper grows rapidly in height to a maximum of 6-8 m above adjacent bare ice. Close to the snout a rapid decline to 3.5 m is evident. The debris supply to the moraine is clearly of englacial origin; individual clasts were observed being exposed at the surface by ablation.
(ii) The eastern moraine also begins as a line of patchy superficial debris, extending downglacier for some 75 m, beyond which the ridge grows rapidly to a maximum height of 14-16 m; there is again a decline immediately above the snout. An important feature of this moraine is that the englacial septum comprises a number of sharply defined transverse debris bands, dipping upglacier at a high angle (70-90°). In all, 28 individual bands were identified in the 'zone of emergence'; these varied in lateral extent from 1 to 11 m, and very rarely exceeded 10-20 cm in thickness. Spacing of the bands varied, but tended to decrease downglacier, 20 bands occurring within 30 m before the commencement of the ridge proper. The debris released from the bands by ablation seems insufficient to give a beaded effect, except in a subdued and rudimentary fashion.

Two questions are posed by the Bas Glacier moraines. First, does the late emergence of the debris, in a zone stretching only some 200 m upglacier from the snout, imply the presence of subglacial debris, carried upwards by shear planes developed in the zone of compressive flow close to the glacier terminus? Such shear planes can be seen on the snout itself, though these stretch across the glacier and do not appear to be concentrating debris along discrete longitudinal zones (as is necessary for medial moraines to form). Alternatively, the debris may be transported within the lower layers of the glacier (the basal zone of transport recognised by Boulton, 1978), although derived from the headwall of the firn zone by supraglacial gelifraction or glacial plucking. Secondly, if the debris within the septa is the result of headwall 'erosion', why are they so narrow? Lateral compression of the glacier tongue may play some part, but it is realistic to suppose that 'point sources' for the debris are involved. Such sources, comprising rock outcrops of limited lateral extent and subject to supraglacial weathering and (possibly) subglacial erosion, can be identified on the upper firn zone of the Bas Glacier d'Arolla. Once incorporated within the glacier, as a result of burial by winter snowfall, the resultant debris would form septa passing deep within the glacier and emerging only at the surface low on the ablation zone, either as a result of ablation itself or as the flow-lines of the glacier turn upwards.

(c) The Haut Glacier d'Arolla

This glacier also supports two medial moraines, though on an altogether larger scale than those of the Bas Glacier.
(i) The western moraine is a classic ISI moraine, commencing at the base of the peak of La Vierge close to the equilibrium

Figure 9.5. Longitudinal debris ridges, marking the outcrop of extended debris bands, at the head of the eastern medial moraine, Haut Glacier d'Arolla.

line of the glacier. It appears to be supplied directly by falls of debris from the flanks of La Vierge directly onto the surfaces of marginal ice-streams. However, clearance of debris from the moraine ridge has revealed longitudinal debris bands, as little as 6-8 cm in width and comprising bubble-free ice containing silt, sand, grit and small angular stones. The overall contribution to the debris cover of this septum seems small, a fact which does much to explain the subsequent morphological evolution of the moraine ridge. The moraine extends some 2.6 km from source to snout; height and width increase over this distance, reaching 11 m and 60.5 m respectively on the last surveyed cross-profile above the glacier snout. Despite its considerable length, the moraine therefore remains relatively poorly developed, reflecting a limited initial supraglacial debris input and an even smaller englacial input via the septum. As a result it displays only the waxing stage of development.

(ii) The eastern moraine is a complex and ultimately more prominent feature, of the AD Above Firn-line Sub-type. Debris begins to appear at the glacier surface only 1 km from the snout, forming initially small but well-defined parallel ridges in the direction of glacier flow (figure 9.5). These merge downglacier to form a single large ridge,

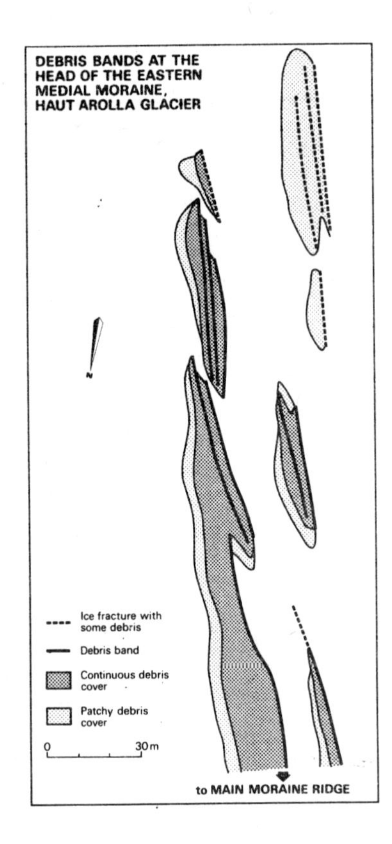

Figure 9.6. Map of debris bands, eastern moraine of the Haut Glacier d'Arolla.

reaching a height of 15-18 m and a width of nearly 100 m. The most notable feature is the emergence of debris not from transverse bands (as in the case of the Bas Glacier d'Arolla), but from longitudinal planes dipping steeply westwards towards the centre-line of the glacier (figure 9.6). Within these planes (which are often lines of fracture) there are large concentrations of mainly coarse debris, including many large slab-like boulders up to 1 m or more along the a axis. The eastern moraine again displays only the waxing stage of development. This reflects the late emergence of a much greater quantity of englacial debris (by comparison with the western moraine). To account for the pronounced and narrow longitudinal septa it is again logical to identify point-sources on the headwall of the accumulation zone. Below these sources the configuration of the basin is such as to produce very powerful lateral compression. The contributing area to the eastern half of the Haut Glacier comprises a broad icefield which by La Vierge is narrowed to 1.1 km and to only 0.4 km where the eastern moraine emerges. A final factor may be the development of large-scale shearing along these compressed debris zones, resulting from the differential velocities between the ice-streams enclosing them.

3. THE FORM AND ORIGIN OF DEBRIS SEPTA IN GLACIERS

(a) Debris septa of the Glacier de Tsidjiore Nouve

In the previous discussion it has been shown that many medial moraines are nourished by englacial debris septa, which may in detail include individual transverse bands (Bas Glacier d'Arolla) or longitudinal bands (Haut Glacier d'Arolla). In 1979 a detailed field examination of the septum at the head of the main moraine of the Glacier de Tsidjiore Nouve was undertaken (Small and Gomez, 1981). As noted above this area is characterised by numerous mounds and ridges; these are actually ice-cored, with a debris cover of variable thickness. Many individual ridges are only some 2 m in length and up to 0.5 m in height; however, where patches of debris have coalesced, irregular and more extensive mounds up to 20 m in width and 4-5 m in height have formed. Downglacier these larger mounds merge to form the prominent medial moraine already described and analysed. All these mounds and ridges, in a zone extending for 360 m upglacier from the moraine ridge, were wholly or partly cleared of overlying debris, revealing the presence of numerous transverse englacial debris bands; all bands were assigned identifying letters. The bands increase in density downglacier; thus only 12 bands (JJ to X) outcrop in the 260 m below the first debris emergence, but 23 bands (W to A) are concentrated in the final 100 m before the moraine head. The debris bands are consistently orientated from south-east to north-west (median bearing 305°), in relation to the centre-line of the glacier (lying 100 m to the west) which runs south-south-west to north-north-east (bearing 10°). The dip of many bands is close to vertical, though some dip at 70-80° upglacier (figure 9.7). There is much variation in thickness both within and between bands. Band P is particularly rich in debris, and reaches a maximum thickness of 30 cm; many other bands lie within the 10-20 cm range. On visual inspection nearly all the bands appear to be overwhelmingly composed of angular debris, including many slab-like clasts with major axes orientated vertically and/or parallel to the margins of the band, within an admixture of gravel and coarse-medium sand. However, bands O and BB are strikingly different, and comprise thin seams of fine sand and silt.

A long-profile along the centre-line of the debris bands outcrop reveals three possible 'waves' on the glacier surface, though these are less prominent than equivalent features at the base of the ice-fall on the Bas Glacier d'Arolla. If the hypotheses of Posamentier (1978) as outlined above were correct, the coincidence between such waves and points of debris emergence should be close. Debris is exposed in considerable amounts at the crest of the lowermost wave (though there are reasons to believe that this wave may be the product of differential ablation rather than compression within the glacier), but not at the next wave upglacier. Moreover, a series of debris bands coincides with the <u>trough</u> between these two waves. The most

Figure 9.7. Debris bands DD and EE, Glacier de Tsidjiore Nouve. Note the steep dip of the bands, and the angular surface boulders which have emerged from them.

noteworthy structural features are narrow bands of compact white ice (comprising fine crystals and numerous air cells) which contrast with the coarsely crystalline ice dominating this zone of the glacier. King and Lewis (1961) argue that such bands represent crevasse fillings of snow, derived from the ice-fall and transformed downglacier. Crevasses developed and infilled within the firn zone are likely to show similar characteristics.

(b) Transport paths of sediments through glaciers.

As a prelude to discussion of the debris bands of the Glacier de Tsidjiore Nouve it is appropriate to consider the question of sediment transport paths through glaciers, as reviewed by Boulton (1978). Such sediment can be derived either from subglacial or supraglacial sources.

(i) Subglacial debris is supplied either directly by the processes of glacial abrasion and plucking or indirectly through basal melting, which releases debris from an englacial to a subglacial position. Subsequently, subglacial debris of either type may be re-incorporated within the glacier by transport along shear planes (for instance in the zone of compression close to the glacier snout) or, possibly, downglacier of a confluence or nunatak where basal flow lines can be expected to transfer material from the glacier base into high-level transport. In temperate glaciers such as those of the Alps, with high rates of basal sliding, debris moved at the base of the ice will be greatly modified by crushing and abrasion, processes which should result in distinctive sedimentological characteristics.

(ii) Supraglacial debris is formed from weathering and collapse of the glacier headwalls and marginal slopes. In Boulton's view, clasts from the valley headwall will be incorporated immediately, forming a zone of low-level englacial debris which follows the buried part of the headwall and the valley floor. Some of this material will become, as a result of progressive basal ice melting, subglacial debris. However, debris from the margins of the glacier within the firn zone will be incorporated, but will travel in 'high-level transport'. This debris will be exposed by ablation immediately below the firn line, and will contribute to the formation of lateral moraines and, where coalescence of ice-streams occurs as in the ISI model, medial moraines. On *a priori* grounds one could infer that the Tsidjiore Nouve moraines are the product of high-level englacial transport of debris (Small and Clark, 1974). In this type of sediment transfer the original characteristics of weathered clasts (frequently blade-shaped and predominantly sharply angular) will be maintained by passive transport within the flowing ice, in contrast to the debris of the 'basal traction zone' which will become spheroidal, comminuted and striated. It should be emphasised that Boulton's distinction between high-level and low-level transport should not be applied too rigidly. In the sense that every part of a glacier is formed originally from snow lying on the glacier surface, and that any part of the glacier surface may receive an increment of debris, supraglacial debris may occur at any level within the ice.

(c) The formation of englacial debris bands.

The segregation of debris into discrete bands within the glacier may occur in four ways (Boulton, 1967).
(i) They may have accumulated in former crevasses, now closed by glacier movement.
(ii) They may represent former surface sedimentary layers, dominantly though not invariably of supraglacial debris, incorporated by successive winter snowfalls on the firn zone.
(iii) They may derive from subglacial debris frozen onto the glacier sole and raised towards the surface as flow-lines turn upwards at subglacial rock projections or near the frontal margins. This mechanism is most appropriate to cold-based glaciers.
(iv) They may, as previously suggested, comprise subglacial debris moved upwards in association with shear planes and major fault-lines in the glacier. From a study of emergent debris bands and associated ice-cored ridges, Boulton infers that this process has operated close to the snout of Sørbreen, Vestspitsbergen. A *prima facie* case for this mechanism contributing to the development of the medial moraines of the Bas Glacier d'Arolla could be made.

(d) Particle-size distributions of englacial debris bands.

An analysis of the sedimentary characteristics of the Tsidjiore Nouve debris bands has been undertaken, on the grounds (stated by Boulton, 1978) that grain-size distributions can provide a good indication of genesis and transport paths of sediment through glaciers. In this study, sediment samples were extracted from 24 of the 35 debris bands at the head of the main moraine. For purposes of comparison samples were also taken from four prominent shear planes outcropping on the terminal slope of the near-by Bas Glacier d'Arolla. The 28 samples were dry sieved in the laboratory, weighed at 0.5ϕ intervals, and calculations made of mean and median particle sizes, skewness, kurtosis and sorting coefficients (figure 9.8). Mean cumulative weight percentages at 0.5ϕ intervals for all debris bands, all shear planes and selected debris bands are shown in figure 9.9. Actual cumulative weight percentage curves for individual debris bands F,M,X,Z and shear planes A,B,C,D are plotted in figure 9.10. From the analysis, the relative coarseness of the debris band particles is immediately evident; the mean grain size of the debris bands is -1.22ϕ (standard deviation .36), and of the shear planes -0.30ϕ. In fact this represents an understatement of true coarseness for the debris bands, for many large clasts - including fragments of up to small boulder size - had to be discarded during the field sampling. Even in the absence of these, the debris bands are dominated by pebbles and granules, while the shear planes contain greater quantities of medium and fine sand and silt. There is some difference in level of sorting between the two sets of debris, with the debris band sediments (sorting coefficient 2.15) being rather better sorted than the shear plane sediments (2.51).

	Mean ϕ	Median ϕ	Sort. Coeff.	Skew.	Kurt.
Debris Bands					
A	-1.38	-1.8	2.25	.43	1.31
C	-1.45	-1.8	2.12	.38	1.28
E	-1.23	-1.7	2.48	.38	1.07
F	-2.81	-3.65	1.91	.61	1.06
H	-1.25	-1.75	2.50	.38	1.13
I	-1.17	-1.65	2.36	.42	1.25
L	-1.30	-1.40	1.83	.26	1.30
M	-1.47	-1.60	1.63	.22	1.26
O	1.32	0.35	3.03	.43	0.90
P	-0.85	-1.35	2.77	.38	1.00
Q	-1.47	-2.05	2.46	.44	1.21
R	-0.38	-1.00	2.45	.38	1.09
T	-1.55	-1.90	2.17	.39	1.30
V	-0.53	-0.90	2.33	.31	1.10
X	-1.68	-1.90	1.90	.23	1.02
Y	-0.81	-1.30	2.21	.42	1.67
Z	-2.28	-2.50	1.40	.34	1.28
BB	-0.15	-0.45	2.52	.22	1.04
CC	-1.23	-1.45	1.60	-24	1.13
DD	-1.46	-1.80	1.94	.35	1.16
EE	-1.58	-1.90	2.13	.38	1.27
GG	-1.45	-1.80	2.19	.37	1.25
HH	-1.45	-1.55	1.55	.18	1.11
JJ	-1.70	-1.95	1.87	.34	1.33
Shear Planes					
A (1)	-1.15	-1.45	2.17	.28	0.82
B (2)	-0.12	-0.60	2.61	.27	0.85
C (3)	-0.48	-0.65	2.55	.13	1.05
D (4)	0.51	0.35	2.71	.60	0.83

Figure 9.8. Grain size analysis of debris bands and shear planes.

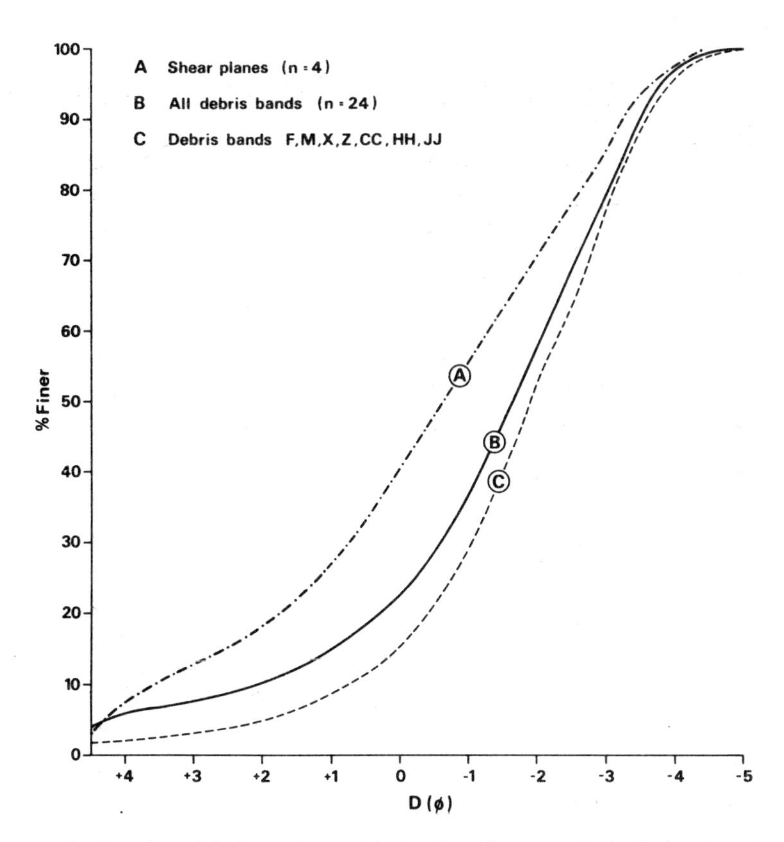

Figure 9.9. Particle size distributions of debris bands and shear planes.

Individual debris bands O and BB require special comment. Mean grain size in each is much reduced (that of O is 1.32ϕ, and that of BB -0.15ϕ), sorting is relatively poor (sorting coefficient are 3.03 and 2.52), and both display coarse-skewed distributions. There is, in fact, a closer resemblance between these debris bands and the shear planes than with the remaining 22 debris bands.

(e) Possible sources of debris for the Glacier de Tsidjiore Nouve.

The presence and characteristics of the Tsidjiore Nouve debris bands pose three interrelated questions. What are the precise sources of the debris? How was the debris incorporated within the glacier? Why is the debris concentrated in discrete transverse bands of limited extent?

Several possible debris sources have been revealed by field examination (figure 9.11).

(i) South of the Col de Tsidjiore Nouve (which links the upper and lower firn basins) a 100 m high rock-face outcrops between flanking ice-streams. The summit of the face is occupied by an ice-cliff which generates small avalanches

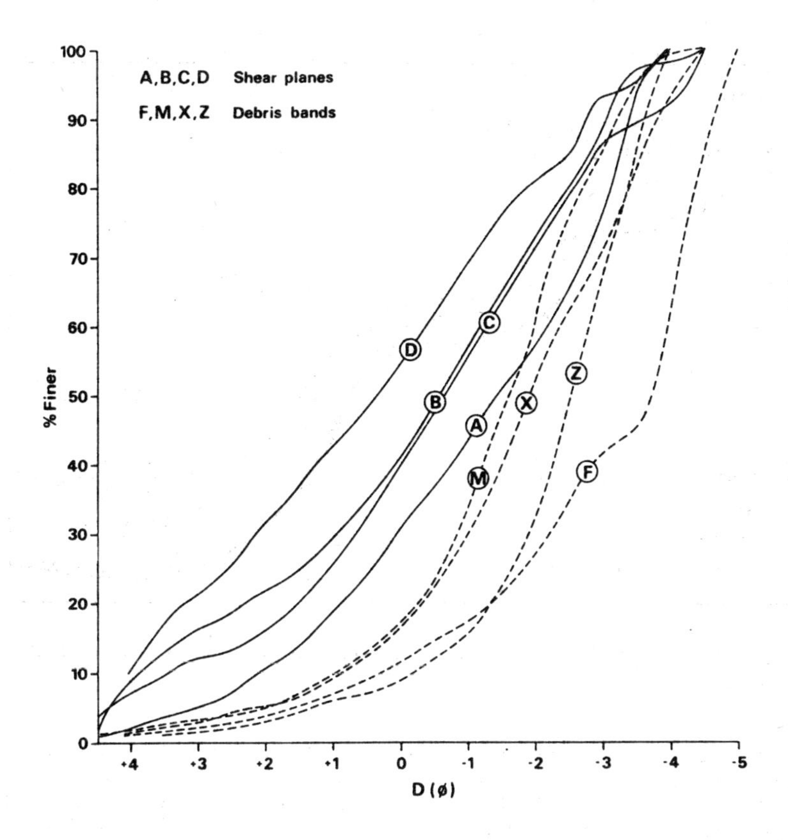

Figure 9.10. Particle size distributions of selected individual debris bands and shear planes.

to the lower firn basin. Beneath the face patches of debris mantle the snow surface; a considerable proportion of this material appears to enter the bergschrund and associated crevasses that occur here. Other similar though smaller rock exposures occupy the north-western slopes of the Pigne d'Arolla, providing more limited debris sources.
(ii) On either flank of the Pigne d'Arolla ice-fall, extensive rock-faces feed large amounts of detritus onto the margins of the glacier. This debris appears to result from subaerial rock disintegration, and provides the debris cover for the prominent lateral moraines which develop on the glacier tongue. These sources are not indicated on figure 9.11.
(iii) Subglacial sources of debris are presumed to exist along the entire length of the glacier. However, areas of concentrated debris release are likely to occur beneath the ice-streams separating the upper and lower firn basins (which, with the exposed rock-face described, constitute a 'headwall' to the lower firn basin), and beneath the Pigne d'Arolla ice-fall where basal velocities are high. Such debris could be moved into high-level transport, and thus

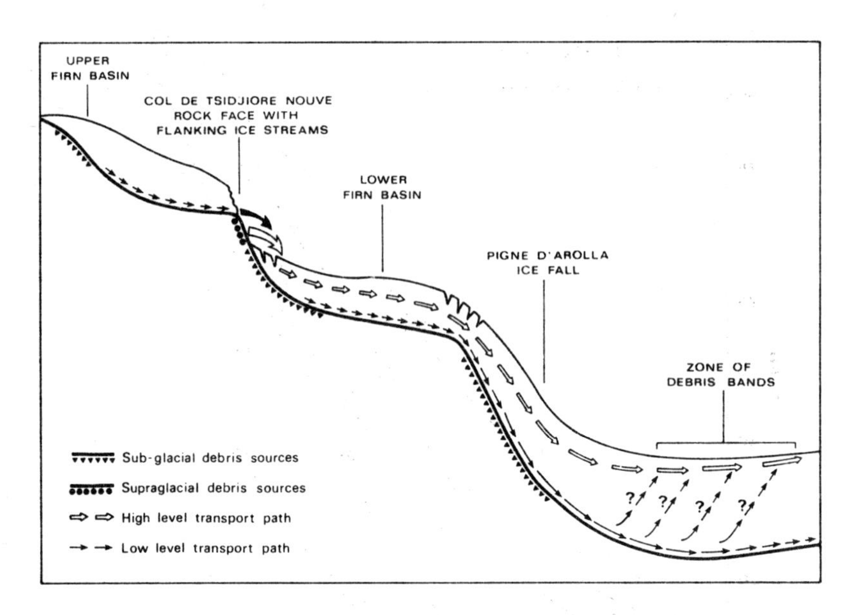

Figure 9.11. Debris sources of the Glacier de Tsidjiore Nouve.

contribute to the medial moraines, only by folds and/or faults at the base of the ice-fall.

As Boulton (1978) has shown, grain-size distributions of debris derived supraglacially from weathering and rock-fall may differ significantly from that transported sub-glacially. The former, if entrained within the glacier as crevasse infillings or sedimentary layers, might experience slight modification (though this would be much less than that produced by crushing, abrasion and attrition beneath the ice) or even undergo purely passive transport. Debris moved along shear planes might be expected to display intermediate degrees of comminution, although it must be stressed that if the planes penetrate to the glacier sole the incorporated debris will be highly comminuted in the first instance.

The particle-size analysis described above has revealed a distinction between the debris band sediments (coarser and predominantly angular) and shear plane sediments (finer and more poorly sorted). The debris band material in fact closely resembles 'moraine debris' (Eyles, 1978), which may either be derived subglacially in the firn zone but be transported passively within the lower part of the glacier or result from supraglacial weathering and incorporation within the firn basin from above. The high-level position of the Tsidjiore Nouve debris supports the latter conclusion, with much of the included debris resulting from rock disintegration of exposed faces between the upper and lower firn basins. The shear plane sediments, by contrast, show characteristics which are typical of 'basal debris' (Eyles, 1978), namely reduced mean grain size, poor sorting and a coarse-skewed distribution resulting from intense comminution. These conclusions are supported by the work of Boulton (1978), whose analysis of high-level transport and zone of traction debris from Breidamerkurjökull and the Glacier d'Argentiére has produced particle size distributions closely similar to those shown in figure 9.9.

Nevertheless, some reservations need to be made. The debris band samples analysed contain a sizeable element of finer material (sometimes up to 10% or more of fine sand, silts and clay). Two (O and BB) are quite distinctive, containing 32% and 20% within this range. One must assume either that the debris bands contain material of supraglacial origin that has not been transported passively within the ice but has locally undergone major alteration, or that the constituents have been mainly derived from supraglacial sources but contain also an element of subglacial origin (which in bands O and BB becomes preponderant). The latter interpretation seems to accord best with the field evidence. As stated, exposed rock-faces that are undoubtedly subjected to mechanical weathering occur within the firn basin of the Glacier de Tsidjiore Nouve. At the same time some subglacial debris may be released from the base of the ice standing above these rock-faces; thus for once such debris may become temporarily supraglacial in position (figure 9.11). Subsequent incorporation of both types of debris on the lower firn basin is most likely to result from ingestion via crevasses (including the bergschrund), a process that is observable in the field. The form, dimensions, disposition and sedimentology of the debris bands studied are generally consistent with the fossil crevasse hypothesis.

4. TYPES OF MEDIAL MORAINE

The primary classification of medial moraines into Ice Stream Interaction (ISI) and Ablation Dominant (AD) types is based more on the origin and emergence of the covering debris than moraine morphology as such. The six medial moraines of the Glacier de Tsidjiore Nouve, Bas Glacier d'Arolla and Haut Glacier d'Arolla illustrate the considerable diversity of form even within one small area. Such diversity reflects, of course, the complex interactions depicted by figure 9.3. In 'ideal' circumstances fully developed medial moraines will comprise waxing, waning and (probably) constant sections. The failure of four of the moraines at Arolla even to approach 'full development' depends on three main factors, all related to debris supply (Small, Clark and Cawse, 1979).

(i) Volume of debris. If this is considerable, differential ablation will cause a sequence of moraine growth, rapid lateral sliding, attenuated debris cover, accelerated ablation and moraine decline (the latter sometimes 'protracted' to give a final constant section). If debris supply is insubstantial, differential ablation will be less marked and the stage of moraine growth itself protracted.

(ii) The balance between direct supraglacial and indirect englacial debris supply. If the moraine is nourished by debris falls within the ablation zone, it will grow immediately and at a rate commensurate with debris increments. If, however, it is supplied by falls in the firn zone, the debris will become englacial and eventually become re-exposed on the ablation zone; if lying well below the glacier surface at the equilibrium line, this debris will reappear 'late' and only the waxing stage can be formed.
(iii) The length of the glacier. This is important in the sense that *ceteris paribus* the longer the glacier the greater the opportunity for full morphological development. However, even on long glaciers the precise point at which debris accumulates on the surface is critical; where the debris is released from englacial septa deep within the glacier, moraine development will be retarded.

The six moraines of the Arolla area appear to fall into three categories:

Type A. This is fed by supraglacial sources close to or below the equilibrium line. The moraine grows in height from source, and if the total debris supply were sufficient a waning section would eventually be initiated. The western moraine of the Haut Glacier d'Arolla is of this type, though it should be noted that there is some englacial feed and that debris amounts are insufficient to result in the waning stage.

Type B. This is fed by falls of rock onto the ice surface well above the equilibrium line, from rock-faces constituting point sources. Debris is incorporated as a series of annual increments, which are buried beneath ice formed lower on the firn zone. On the ablation zone, the englacial debris will melt out to give a relatively short moraine, which may be dominated by the waxing stage. The eastern moraine of the Haut Glacier and the western of the Bas Glacier are probably of this type, though the debris septa of the latter have been greatly modified by lateral compression and, possibly, longitudinal shearing.

Type C. This is also fed above the equilibrium line, but debris incorporation is by way of crevasses forming adjacent to the point sources. The resultant steep debris bands melt out on the ablation zone; where the bands are rich in sediment beaded medial moraines form. Whether such moraines develop into prominent ridges (via the waxing stage) and then undergo decline will depend on (i) total debris inputs, and (ii) the precise point of emergence. On the Glacier de Tsidjiore Nouve both factors are favourable; on the eastern moraine of the Bas Glacier d'Arolla they are not.

REFERENCES

1. Boulton, G.S., 1967, The development of a complex supraglacial moraine at the margin of Sørbreen, Ny Friesland, Vestspitsbergen, *Journal of Glaciology, 6,* 717-735.

2. Boulton, G.S., 1978, Boulder shapes and grain-size distributions of debris as indicators of transport paths through a glacier and till genesis, *Sedimentology, 25*, 773-799.

3. Embleton, C. and King, C.A.M., 1968, *Glacial and Periglacial Geomorphology,* (Edward Arnold (Publishers) Ltd, London).

4. Eyles, N. and Rogerson, R.J., 1978, A framework for the investigation of medial moraine formation; Austerdalsbreen, Norway, and Berendon Glacier, British Columbia, Canada, *Journal of Glaciology, 20,* 93-113.

5. Eyles, N., 1978, Scanning electron microscopy and particle size analysis of debris from a British Columbian glacier: a comparative report, in: *Scanning Electron Microscopy in the study of sediments: a symposium,* ed. Whalley, W.B., (Geo Abstracts, Norwich), 227-241.

6. King, C.A.M. and Lewis, W.V., 1961, A tentative theory of ogive formation, *Journal of Glaciology, 3,* 913-939.

7. Loomis, S.R., 1970, Morphology and ablation processes on glacier ice, in: *Icefield Ranges Research Project. Scientific Results,* ed. Bushnell, V.C. and Ragle, R.H. New York, American Geographical Society; Montreal, Arctic Institute of North America, 2, 27-31.

8. Østrem, G., 1959, Ice melting under a thin layer of moraine, and the existence of ice cores in moraine ridges, *Geografiska Annaler, 4,* 228-230.

9. Posamentier, H., 1978, Thoughts on ogive formation, *Journal of Glaciology, 82,* 218-230.

10. Sharp, R.P., 1948, The constitution of valley glaciers, *Journal of Glaciology, 1,* 174-175, 182-189.

11. Sharp, R.P., 1949, Studies of superglacial debris on valley glaciers, *American Journal of Science, 247*, 289-315.

12. Small, R.J. and Clark, M.J., 1974, The medial moraines of the lower Glacier de Tsidjiore Nouve, Valais, Switzerland, *Journal of Glaciology, 13*, 255-263.

13. Small, R.J. and Clark, M.J., 1976, Morphology and development of medial moraines: a reply to comments by N. Eyles, *Journal of Glaciology, 75*, 162-164.

14. Small, R.J., Clark, M.J. and Cawse, T.J.P., 1979, The formation of medial moraines on Alpine glaciers, *Journal of Glaciology, 22,* 43-52.

15. Small, R.J. and Gomez, B., 1981, The nature and origin of debris layers within the Glacier de Tsidjiore Nouve, Valais, Switzerland, *Annals of Glaciology, 2,* 109-113.

110234

LINKED

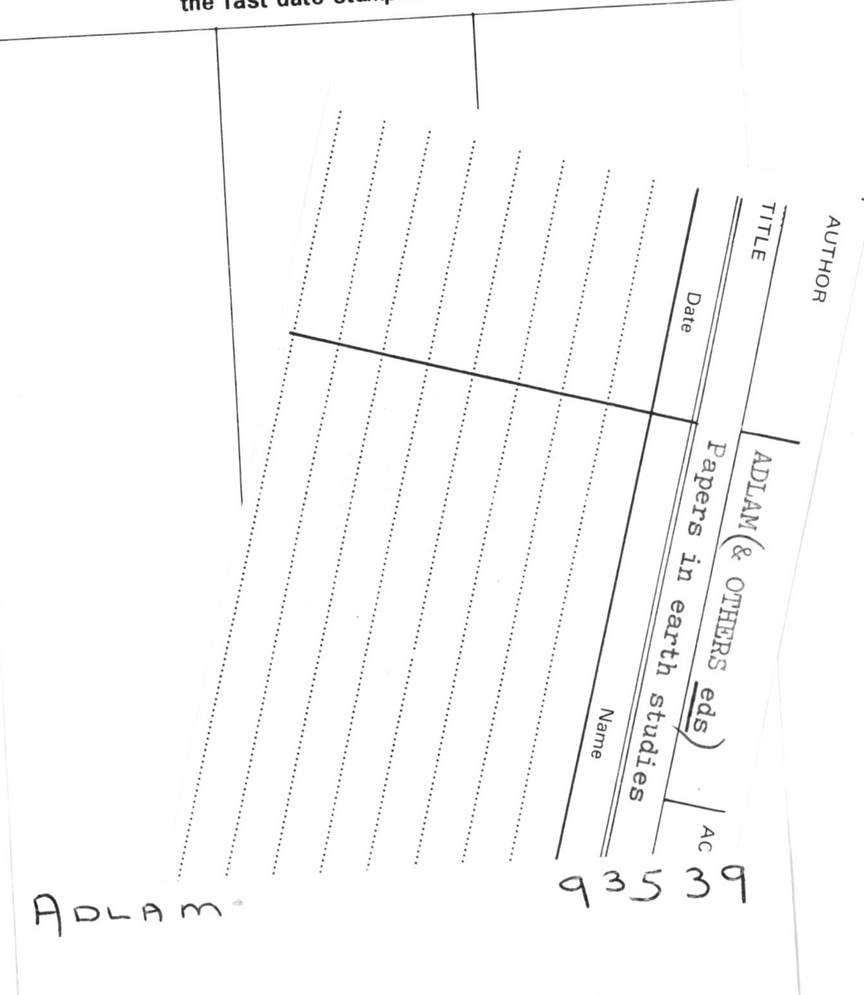
AUTHOR
TITLE
Date
Name
ADLAM (& OTHERS eds)
Papers in earth studies
93539